A Practical Guide to the Feeding of Organic Farm Animals

A Practical Guide to the Feeding of Organic Farm Animals

Pigs, Poultry, Cattle, Sheep and Goats

Robert Blair

5m Publishing

First published 2016

Published by
5M Publishing Ltd,
Benchmark House,
8 Smithy Wood Drive,
Sheffield, S35 1QN, UK
Tel: +44 (0) 1234 81 81 80
www.5mpublishing.com

A Catalogue record for this book is available from the British Library

ISBN 978-1-910455-70-8

Book layout by Servis Filmsetting Ltd, Stockport, Cheshire
Printed by Replika Press Pvt Ltd, India
Photos by Robert Blair unless otherwise credited

Robert Blair
Faculty of Land and Food Systems
University of British Columbia
Vancouver, British Columbia
Canada

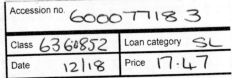

Contents

CHAPTER 1

Introduction

Feed represents a major portion of the costs of raising organic stock, especially since organic feedstuffs are generally more expensive and in shorter supply than conventional feedstuffs. Therefore, the profitability of organic milk, meat and egg production requires that the dietary mixtures are formulated correctly so that the stock grows and produces milk, meat or eggs efficiently. Basic to the selection of feedstuffs for each class of farm stock (poultry, pigs, cattle and sheep) is an understanding of what nutrients are required in their respective diets and the extent to which the animal or bird in question can digest the feed to obtain these nutrients.

Another objective with organic feed is to ensure that the diet does not contain imbalances or excesses of nutrients such as protein and phosphorus, resulting in breakdown products that are excreted in the manure and cause environmental damage.

There is field evidence that some of the organic feed being used on farms does not meet the required quality standards. It is to be hoped that this book will help to prevent that situation.

Organic regulations pose challenges and problems for the feed formulator, in part due to a lack of detail in the standards. For instance, as noted by Wilson (2003), the UK regulations prohibit materials produced with the use of 'genetically modified

organisms or products derived therefrom'. A problem raised by this definition is how far down the production chain the prohibition applies. For instance, vitamin B2 and vitamin B12 are generally produced using a fermentation process and in the case of vitamin B12 the organism used predominantly is a genetically modified (GM) strain. A strict interpretation of the regulations would therefore exclude this vitamin, which would have to be provided by the main ingredients. Unfortunately, this vitamin is absent from grains and plant materials and occurs only in ingredients of animal origin. Another example is that Sweden has approved the growing of GM potatoes for the production of starch for use in the paper industry. This has resulted in a useful by-product, potato protein, being available for use in animal feed since Sweden is one of the countries facing an extreme shortage of organic protein feeds, exacerbated by the ban on pure amino acids. Is potato protein derived from GM potatoes acceptable in organic feed? The answer depends on how the organic regulations are interpreted.

Wilson (2003) also pointed out the omission in the EU regulations (European Commission, 1999) of extracted oil from the list of permitted ingredients, although oilseeds and their by-products are permitted. This omission is difficult to understand.

A possible explanation is that the EU regulations assume that extracted oil is used exclusively for the human market. The list of permitted ingredients in New Zealand (NZFSA, 2005), which appears to be based on the EU list, clarifies this issue by permitting plant oils obtained from approved oilseeds by mechanical extraction. The fact that maize as bran is listed twice in the EU list of permitted grains suggests that the omission of extracted oil may simply be a clerical error. These examples highlight the need for detailed specifications in the organic regulations and for an enlightened approach by certifying agencies in their interpretation.

Fishmeal is allowed in organic feed as a good source of amino acids but there is little fishmeal available that does not contain the synthetic preservative ethoxyquin, which is not permitted in organic production. This indicates that farmers have to ask quite detailed and knowledgeable questions when purchasing organic feedstuffs.

Although the main aim of this book is to assist organic producers in formulating diets and feeding programmes for organic livestock and poultry, the regulatory authorities in several countries may find it of value in addressing nutritional issues relevant to future revisions of the regulations. It seems clear that the current standards and regulations have been developed mainly by those experienced in crop production and in ecological issues and that a review of the regulations from an animal nutrition perspective is required.

REFERENCES

European Commission (1999). Council Regulation (EC) No 1804/1999 of 19 July 1999 supplementing Regulation (EEC) No. 2092/91 on organic production of agricultural products and indications referring thereto on agricultural products and foodstuffs to include livestock production. Official Journal of the European Communities 2.8.1999, L222, 1–28.

NZFSA (2005). NZFSA Technical Rules for Organic Production, Version 5. New Zealand Food Safety Authority, Wellington.

Wilson, S. (2003). Feeding animals organically – the practicalities of supplying organic animal feed. In: Garnsworthy, P.C. and Wiseman, J. (eds) Recent Advances in Animal Nutrition. University of Nottingham Press, Nottingham, UK, pp. 161–172.

CHAPTER 2

Feedstuffs for Organic Feeding

Only those feedstuffs meeting organic standards are permitted in organic diets. No genetically modified (GM) types are allowed. Also, the organic standards are designed to require that most of the feedstuffs needed are produced on-farm.

New Zealand is one of the few countries to include a detailed list of approved feed ingredients in the organic regulations (Table 2.1). This is a very useful feature of its regulations. In addition, the regulations stipulate that the feeds must meet the Agricultural Compounds and Veterinary Medicines (ACVM) Act and regulations, and the Hazardous Substances and New Organisms (HSNO) Act, or are exempt, thus providing additional assurance to the consumer. This list appears to be based

on the EU list, possibly because of export requirements.

The EU has a somewhat similar list (Table 2.1), but one detailing non-organic feedstuffs that can be used in limited quantities in organic feeds. It may be inferred from the EU list that organic sources of the named ingredients are acceptable. This list is particularly useful because it is currently very difficult to formulate some feeds that are 100% organic. As a result the regulations in several regions allow for feed to contain up to 5% non-organic ingredients. Should a producer need to take advantage of this provision, it is necessary that approval be obtained from the local certifying agency. The EU list can then be used to select the appropriate ingredients to make up the 5%.

TABLE 2.1 Comparison of approved organic feedstuffs in New Zealand and approved non-organic feedstuffs in the EU.

	New Zealand approved list (only those named in each category) MAF Standard OP3, Appendix Two, 2011 (NZFSA, 2011)	EU-approved list of non-organic feedstuffs (up to defined limits) Council Regulation EC No 834/2007, 2007 (EU, 2007)
1. Feed materials of plant origin	1.1 Cereals, grains, their products and by-products: oats as grains, flakes, middlings, hulls and bran; barley as grains, protein and middlings; rice germ expeller; millet as grains; rye as grains and middlings; sorghum as grains; wheat as grains, middlings,	1.1 Cereals, grains, their products and by-products: oats as grains, flakes, middlings, hulls and bran; barley as grains, protein and middlings; rice as grains, rice broken, bran and germ expeller; millet as grains; rye as grains, middlings, feed and bran; sorghum

	New Zealand approved list (only those named in each category) MAF Standard OP3, Appendix Two, 2011 (NZFSA, 2011)	EU-approved list of non-organic feedstuffs (up to defined limits) Council Regulation EC No 834/2007, 2007 (EU, 2007)
1. Feed materials of plant origin	bran, gluten feed, gluten and germ; spelt as grains; triticale as grains; maize as grains, bran, middlings, germ expeller and gluten; malt culms; brewers' grains (rice as grain, rice broken, rice bran, rye feed, rye bran and tapioca were delisted in 2004.)	as grains; wheat as grains, middlings, bran, gluten feed, gluten and germ; spelt as grains; triticale as grains; maize as grains, bran, middlings, germ expeller and gluten; malt culms; brewers' grains.
	1.2 Oilseeds, oil fruits, their products and by-products: rapeseed, expeller and hulls; soya bean as bean, toasted, expeller and hulls; sunflower seed as seed and expeller; cotton as seed and seed expeller; linseed as seed and expeller; sesame seed as expeller; palm kernels as expeller; pumpkin seed as expeller; olives, olive pulp; vegetable oils (from physical extraction). (Turnip rapeseed expeller was delisted in 2004.)	1.2 Oilseeds, oil fruits, their products and by-products: rapeseed, expeller and hulls; soya bean as bean, toasted, expeller and hulls; sunflower seed as seed and expeller; cotton as seed and seed expeller; linseed as seed and expeller; sesame seed as seed and expeller; palm kernels as expeller; turnip rapeseed as expeller and hulls; pumpkin seed as expeller; olive pulp (from physical extraction of olives).
	1.3 Legume seeds, their product and by-products: chickpeas as seeds, middlings and bran; ervil as seeds, middlings and bran; chickling vetch as seeds submitted to heat treatment, middlings and bran; peas as seeds, middlings and bran; broad beans as seeds, middlings and bran; horse beans as seeds, middlings and bran; vetches as seeds, middlings and bran and lupin as seeds, middlings and bran.	1.3 Legume seeds, their product and by-products: chickpeas as seeds; ervil as seeds; chickling vetch as seeds submitted to an appropriate heat treatment; peas as seeds, middlings and bran; broad beans as seeds, middlings and bran; horse beans as seeds, vetches as seeds and lupin as seeds.
	1.4 Tuber roots, their products and by-products: sugar beet pulp, potato, sweet potato as tuber, potato pulp (by-product of the extraction of potato starch), potato starch, potato protein and manioc (cassava).	1.4 Tuber roots, their products and by-products: Sugar beet pulp, dried beet, potato, sweet potato as tuber, manioc as roots, potato pulp (by-product of the extraction of potato starch), potato starch, potato protein and tapioca.
	1.5 Other seeds and fruits, their products and by-products: carob, carob pods and meals thereof, pumpkins, citrus pulp, apples, quinces, pears, peaches, figs, grapes and pulps thereof; chestnuts, walnut expeller, hazelnut expeller; cocoa husks and expeller; acorns.	1.5 Other seeds and fruits, their products and by-products: carob pods, citrus pulp, apple pomace, tomato pulp and grape pulp.
	1.6 Forages and roughages: lucerne, lucerne meal, clover, clover meal, grass (obtained from forage plants), grass meal, hay, silage, straw of cereals and root vegetables for foraging.	1.6 Forages and roughages: lucerne, lucerne meal, clover, clover meal, grass (obtained from forage plants), grass meal, hay, silage, straw of cereals and root vegetables for foraging.

	New Zealand approved list (only those named in each category) MAF Standard OP3, Appendix Two, 2011 (NZFSA, 2011)	EU-approved list of non-organic feedstuffs (up to defined limits) Council Regulation EC No 834/2007, 2007 (EU, 2007)
	1.7 Other plants, their products and by-products: molasses, seaweed meal (obtained by drying and crushing seaweed and washed to reduce iodine content), powders and extracts of plants, plant protein extracts (solely provided to young animals), spices and herbs.	1.7 Other plants, their products and by-products: molasses as a binding agent in compound feeding stuffs, seaweed meal (obtained by drying and crushing seaweed and washed to reduce iodine content), powders and extracts of plants, plant protein extracts (solely provided to young animals), spices and herbs.
2. Feed materials of animal origin	2.1 Milk and milk products: raw milk, milk powder, skim milk, skimmed milk powder, buttermilk, buttermilk powder, whey, whey powder, whey powder low in sugar, whey protein powder (extracted by physical treatment), casein powder, lactose powder, curd and sour milk.	2.1 Milk and milk products: raw milks (as defined in Article 2 of Directive 92/46/EEC), milk powder, skimmed milk, skimmed milk powder, buttermilk, buttermilk powder, whey, whey powder, whey powder low in sugar, whey protein powder (extracted by physical treatment), casein powder and lactose powder.
	2.2 Fish, other marine animals, their products and by-products: fish, fish oil and cod-liver oil not refined; fish molluscan or crustacean autolysates, hydrolysate and proteolysates obtained by an enzyme action, whether or not in soluble form, solely provided to young animals. Fishmeal.	2.2 Fish, other marine animals, their products and by-products: fish, fish oil and cod-liver oil not refined; fish molluscan or crustacean autolysates, hydrolysate and proteolysates obtained by an enzyme action, whether or not in soluble form, solely provided to young animals. Fishmeal.
3. Feed materials of mineral origin	3.1 Sodium products: unrefined sea salt, coarse rock salt, sodium sulphate, sodium carbonate, sodium bicarbonate, sodium chloride.	3.1 Sodium products: unrefined sea salt, coarse rock salt, sodium sulphate, sodium carbonate, sodium bicarbonate, sodium chloride.
	3.2 Calcium products: lithotamnion and maerl shells of aquatic animals (including cuttlefish bones), calcium carbonate, calcium lactate, calcium gluconate.	3.2 Calcium products: lithotamnion and maerl shells of aquatic animals (including cuttlefish bones), calcium carbonate, calcium lactate, calcium gluconate.
	3.3 Phosphorus products: bone dicalcium phosphate precipitate, defluorinated dicalcium phosphate, defluorinated monocalcium phosphate.	3.3 Phosphorus products: bone dicalcium phosphate precipitate, defluorinated dicalcium phosphate, defluorinated monocalcium phosphate.
	3.4 Magnesium products: magnesium sulphate, magnesium chloride, magnesium carbonate, magnesium oxide (anhydrous magnesia).	3.4 Magnesium products: anhydrous magnesia, magnesium sulphate, magnesium chloride, magnesium carbonate.
	3.5 Sulphur products: sodium sulphate.	3.5 Sulphur products: sodium sulphate.
Feed additives	Trace elements.	Trace elements.

	New Zealand approved list (only those named in each category) MAF Standard OP3, Appendix Two, 2011 (NZFSA, 2011)	EU-approved list of non-organic feedstuffs (up to defined limits) Council Regulation EC No 834/2007, 2007 (EU, 2007)
	E1 Iron products: ferrous carbonate, ferrous sulphate monohydrate, ferric oxide.	E1 Iron products: ferrous carbonate, sulphate monohydrate and/or heptahydrate, ferric oxide.
	E2 Iodine products: calcium iodate, anhydrous calcium iodate hexahydrate, potassium iodide.	E2 Iodine products: calcium iodate, anhydrous calcium iodate, hexahydrate, sodium iodide.
	E3 Cobalt products: cobaltous sulphate monohydrate and/or heptahydrate, basic cobaltous carbonate monohydrate.	E3 Cobalt products: cobaltous sulphate monohydrate and/or heptahydrate, basic cobaltous carbonate monohydrate.
	E4 Copper products: copper oxide, basic copper carbonate monohydrate, copper sulphate pentahydrate.	E4 Copper products: copper oxide, basic copper carbonate monohydrate, copper sulphate pentahydrate.
	E5 Manganese products: manganous carbonate, manganous oxide, manganous sulphate, monohydrate and/or tetrahydrate.	E5 Manganese products: manganous carbonate, manganous oxide, manganous sulphate monohydrate and/or tetrahydrate.
	E6 Zinc products: zinc carbonate, zinc oxide, zinc sulphate monohydrate and/or heptahydrate.	E6 Zinc products: zinc carbonate, zinc oxide, zinc sulphate monohydrate and/or heptahydrate.
	E7 Molybdenum products: ammonium molybdate, sodium molybdate.	E7 Molybdenum products: ammonium molybdate, sodium molybdate.
	E8 Selenium products: sodium selenate, sodium selenite.	E8 Selenium products: sodium selenate, sodium selenite.
Vitamins and provitamins	Vitamins approved for use under NZ legislation: preferably derived from ingredients occurring naturally in feeds, or synthetic vitamins identical to natural vitamins only for non-ruminant animals. When the organic feed or organic pork product is to be exported to the USA, the vitamins and trace minerals used have to be FDA-approved.	Vitamins, provitamins and chemically well-defined substances having a similar effect. Vitamins authorized under Directive 70/524/EEC: preferably derived from raw materials occurring naturally in feedstuffs, or synthetic vitamins identical to natural vitamins (only for non-ruminant animals).
Enzymes	Enzymes approved for use under NZ legislation.	Enzymes authorized under EU Directive 70/524/EEC.
Microorganisms	Microorganisms approved for use under NZ legislation.	Microorganisms authorized under EU Directive 70/524/EEC.
Preservatives	E 236 Formic acid.	E 236 Formic acid only for silage.
	E 260 Acetic acid.	E 260 Acetic acid only for silage.
	E 270 Lactic acid.	E 270 Lactic acid only for silage.
	E 280 Propionic acid.	E 280 Propionic acid only for silage.

	New Zealand approved list (only those named in each category) MAF Standard OP3, Appendix Two, 2011 (NZFSA, 2011)	EU-approved list of non-organic feedstuffs (up to defined limits) Council Regulation EC No 834/2007, 2007 (EU, 2007)
Binders, anticaking agents and coagulants	E 551b Colloidal silica.	E 551b Colloidal silica.
	E 551c Kieselgur.	E 551c Kieselgur.
	E 558 Bentonite.	E 553 Sepiolite.
	E 559 Kaolinitic clays.	E 558 Bentonite.
	E 561 Vermiculite.	E 559 Kaolinitic clays.
	E 562 Sepiolite.	E 561 Vermiculite.
	E 599 Perlite.	E 599 Perlite.
Antioxidant substances	E 306 Tocopherol-rich extracts of natural origin.	E 306 Tocopherol-rich extracts of natural origin.
Certain products used in animal nutrition	Brewer's yeast.	Brewer's yeast.
Processing aids for silage	Sea salt, coarse rock salt, whey, sugar, sugar beet pulp, cereal flour and molasses.	Sea salt, coarse rock salt, enzymes, yeasts, whey, sugar, sugar beet pulp, cereal flour, molasses and lactic, acetic, formic and propionic bacteria.
Pure amino acids	None.	None.

Most countries follow the EU system and do not publish an approved list of organic feedstuffs, stating that all feedstuffs used must meet organic guidelines. An example is the USA, where the regulations also state that all feed, feed additives and feed supplements must comply with Food and Drug Administration (FDA) regulations. Canada has a much less detailed Permitted Substances List (CAN/CGSB-32.311-2006) than the New Zealand or EU lists, stating that "… energy feeds and forage concentrates (grains) and roughages (hay, silage, fodder, straw) shall be obtained from organic sources and may include silage preservation products (e.g. bacterial or enzymatic additives derived from bacteria, fungi and plants and food by-products [e.g. molasses and whey]). Note that if weather conditions are unfavourable to

fermentation, lactic, propionic and formic acid may be used."

It would be very helpful to organic feeders if more countries followed the New Zealand and EU models and provided detailed lists.

Although the New Zealand and EU regulations state that pure amino acids such as lysine and methionine are not allowed in organic feed, the regulations in several regions allow a temporary usage of these feed additives to improve the quality of the protein. For instance, Canada has banned the use of amino acids from synthetic sources but has granted an exception for the use of synthetic DL-methionine, DL-methionine-hydroxy analog and DL-methionine-hydroxy analog calcium, with this exception to be re-evaluated at the next revision of the standard. Currently the

Canadian provinces of Quebec and British Columbia allow pure amino acids, a distinction being made between those of fermented origin (approved, e.g. lysine) and those of synthetic origin (prohibited, e.g. methionine, except for poultry feed).

The information in the above table, which is drawn from both the northern and southern hemispheres, can be used as a potential list of the feedstuffs available for organic feeding in many countries. Not all the feed ingredients in Table 2.1 are suitable for inclusion in all types of diets since, for example, some are more suited to cattle feeding than poultry and pig feeding. In addition, some of the ingredients may not be grown on-farm or usually available in sufficient quantity.

One of the questions raised by the publication of lists of approved feed ingredients and supplements in organic regulations is how new ingredients are added. An example is insect larval meal, which is being produced in several countries from substrates such as household food waste. Some certifying agencies take the view that the meal produced is a permissible ingredient (and a good source of protein) in organic feed, while others take the view that the insect meal is unacceptable as an organic feed ingredient since the substrate used cannot be verified as organic. Producers planning to use this type of product should, therefore, obtain approval from a local certifying agency before doing so.

CEREAL GRAINS AND BY-PRODUCTS

The primary sources of energy for pig and poultry feeding are cereal grains. Also, the processing of cereals for the human market yields by-products that are important as feed ingredients. In general, cereal grains are low

in protein but high in carbohydrate and they are generally palatable and well-digested when passed through a grinder before being included in the diet. Nutrient composition can be quite variable.

Organic cereals generally have a lower content of protein, due to fertilizer practices and production methods, but in other respects are similar to conventional cereals in nutrient content (Blair, 2012). Some animals appear to have the ability to discriminate between organic and conventional sources of cereals. In preference tests, rats selected biscuits made from organic wheat over biscuits made from conventional wheat. Conversely, wild birds preferred conventional rather than organic wheat as a winter feed, a result ascribed to a lower content of protein in the organic wheat.

The fibre in grains is contained mainly in the hull (husk) and can be variable in content, depending on the growing and harvesting conditions. This can affect the starch content of the seed and, as a consequence, the energy value. The hull is quite resistant to digestion and also has a lowering effect on the digestibility of nutrients.

Yellow maize is the only cereal grain to contain vitamin A, owing to the presence of provitamins (mainly β-carotene). All grains are deficient in vitamin D.

The oil in cereal grains is contained in the germ and varies from less than 20g/kg (dry basis) in wheat to more than 50g/kg in oats.

It is high in oleic and linoleic acids, which are unstable after the seed is ground. As a result, rancidity can develop quickly and result in reduced palatability of the feed or feed refusal.

Feed grains can only meet part of the requirement for dietary nutrients, and other feed components are needed to balance the diet completely. Combining grains and other ingredients into a final dietary mixture to meet the nutritional needs of animals or birds requires information about the nutrient content of each feedstuff and its suitability as a feed ingredient.

Maize, wheat, oats, barley and sorghum are the principal cereals used in organic feeding, as whole grains and/or grain by-products. Generally, maize and wheat are highest in energy value, with sorghum, barley, oats and rye being lower. Some rye is also used. Although it is similar to wheat in composition, it is less palatable and may contain ergot, a toxic fungus. Triticale, a hybrid of rye and wheat, is being used in poultry and livestock feeding in some countries.

There do not appear to be any genetically modified (GM) varieties of wheat, sorghum, barley or oats being grown commercially, unlike the situation with maize. In the USA and some other countries, for instance, substantial quantities of GM maize varieties developed with insect and herbicide resistance are being grown.

Organic grains tend to be variable in nutritional composition due to the fertilizer used and other factors. Therefore it is recommended that a chemical analysis is conducted prior to feeding to determine more exactly their nutrient composition and quality.

Barley

This has been used traditionally as a principal grain for pig and poultry feeding in the western areas of North America, the UK and many countries of Europe. It is considered a medium-energy grain, lower in energy than maize and needs to be ground before use. If no grinding facilities are available, soaking in water for at least 24 hours is an alternative.

Barley contains more fibre than maize and less than oats, but the proportion of hull to kernel is variable, resulting in a variable energy value. The protein content is higher than in maize and can range from about 90 to 160g/kg.

Barley contains more phosphorus than other common cereal grains.

Hull-less barley varieties have been developed in which the hull separates during threshing. These varieties contain more protein and less fibre than conventional barley, and theoretically should be superior in nutritional value to conventional barley. However, several studies have failed to show any superiority, suggesting that hull-less barley should simply be used in organic diets as a substitute for conventional barley on a 1:1 basis.

Barley by-products from distilling and brewing may be used in organic poultry and livestock feeding, but should be verified with the local certifying agency.

Wheat

This cereal is widely cultivated in temperate countries and in cooler regions of tropical countries.

It is commonly used for feeding to pigs and poultry when it is surplus to human requirements or is considered not suitable for the human food market. Otherwise it may be too expensive for organic feeding. However some wheat is grown for feed purposes. By-products of the flour milling industry also are very desirable ingredients in organic diets.

One concern about wheat is that the energy and protein contents are more variable than in other cereal grains such as maize, sorghum and barley. Therefore periodic testing of batches of wheat for nutrient content is recommended. When this is done a wide range of wheat types can be utilized efficiently in organic diets.

During threshing the husk separates from the grain, leaving a less fibrous product. As a result wheat is similar to maize in energy value but contains more protein. Therefore, it can be used as a replacement for maize as a high-energy ingredient and it requires less protein supplementation than maize.

High-moisture wheat preserved by addition of propionic acid can be used, but since the acid destroys vitamin E and some other vitamins care should be taken to ensure that the diet is supplemented adequately.

Wheat is a good ingredient to use in pelleted diets due to its gluten content, which helps to bind the ground feed particles together.

Milling by-products from flour production can also be used in organic feeding. These by-products are usually classified according to their protein and fibre contents and are traded under a variety of names such as pollards, offals, shorts, wheatfeed and midds (wheat middlings). In general, those by-products with low levels of fibre are of higher nutritive value for organic feeding.

Wheat Middlings

The name 'middlings' derives from the fact that this by-product is somewhere in the middle between flour on one hand and bran on the other. This by-product is known as wheatfeed or pollards in Europe and Australia. The composition and quality of middlings vary greatly due to the proportions of fractions included, also the amount of screenings added and the fineness of grind.

When middlings (or whole wheat) are included in pelleted feeds, the pellets are more cohesive and there is less breakage and fewer fines.

Spelt

Spelt is a subspecies of wheat grown widely in Central Europe. It has been introduced to other countries partly for the human market because of its reputation as being low in gliadin, the gluten fraction implicated in coeliac disease.

This crop appears to be generally more winter-hardy than soft red winter wheat, but less winter-hardy than hard red winter wheat. The yield is generally lower than that of wheat.

Recent studies suggest that this grain is similar to wheat in nutritive value, but with a higher protein content. Tests should be carried out to confirm the nutritive content before it is used. Expansion of this crop is expected in Europe because of the current shortage of high-protein organic feedstuffs.

Oats

Oats are higher in hull, fibre and ash contents and lower in starch than maize, grain sorghum or wheat. As a result this grain has a much lower energy value than other main cereals. Oats vary in protein content from about 110 to 170g/kg. They have a higher oil content than maize, but this does not compensate for the high fibre content.

Oats are commonly processed mechanically to remove the hull from the kernel. The result is oat groats. After rolling, the product is known as rolled oat groats. This is a highly palatable, high-energy feedstuff. The protein content is around 160g/kg. However, this product is too expensive for general use in organic diets and is used mainly in creep feed and pig starter diets.

Maize

Maize is high in starch and in oil and low in fibre, giving it a high energy value. Other grains, except wheat, have a lower energy value than maize. Maize oil has a high proportion of unsaturated fatty acids and is an excellent source of linoleic acid.

The protein content is low at about 85g/kg and is not well balanced in amino acid content.

Maize is very low in calcium (about 0.2g/kg). It contains a higher level of phosphorus (2.5–3.0g/kg) but much of the phosphorus is bound in phytate form that it is poorly available to pigs and poultry. As a result, a high proportion of the phosphorus passes through the gut and is excreted in the manure. The diet can be supplemented with phytase enzyme to improve phosphorus utilization. Another approach is to use one of the newer low-phytic-acid maize varieties that have about 35% of their phosphorus bound in phytate compared with 70% for conventional maize. These varieties allow the phosphorus to be utilized more effectively so that less is excreted in the manure. As with other improved varieties of maize, producers wishing to use such varieties should check their acceptability with the local organic certifying agency.

The quality of maize is excellent when harvested and stored under good conditions, including proper drying to 100–120g moisture/kg. Fungal toxins can develop in grain that is harvested damp or allowed to become damp during storage, causing health problems particularly in pigs and ducks.

Sorghum

Sorghum, commonly called grain sorghum or milo, is the fifth most important cereal crop grown in the world. Much of it is used in the human market. As a continent, Africa is the largest producer of sorghum. Other leading producers include India, Mexico, Australia and Argentina. Sorghum is one of the most drought-tolerant cereal crops and is more suited than maize to harsh weather conditions such as high temperatures and less consistent rainfall.

Grain sorghum is generally higher in protein than maize and is similar in energy content. However, one disadvantage of grain sorghum is that it is that it can be more variable in composition because of growing conditions. Protein content usually averages around 89g/kg, but can vary from 70 to 130g/kg.

Grain sorghum can be used in a range of organic diets as a replacement for maize. The hybrid yellow-endosperm varieties are more palatable to pigs and poultry and are better utilized than the darker brown sorghums, which have been developed with a higher tannin content to deter wild birds from damaging the crop.

Proper grinding of grain sorghum is important because of the hard seed coat.

Triticale

This cereal is a hybrid of wheat and rye which grows well on poor soils. It has a similar composition to wheat, and is usually slightly higher in important amino acids. Triticale is grown mainly in Poland, China, Russia, Germany, Australia and France, although some is grown in North and South America. It is reported to grow well in regions not suitable for maize or wheat.

The greatest potential for its use as a grain feedstuff is in pig and poultry operations that grow at least part of their own feed grain supply, using lands that are heavily manured. Triticale production in such conditions is generally more productive and sustainable than barley or other cereals for feed grain. Its greater disease resistance compared with wheat and barley is another advantage. Thus, triticale is of particular interest to organic pig and poultry producers.

The newly developed Canadian varieties possess more of the characteristics of the wheat parent than the rye parent, resulting in higher energy content and improved palatability and nutritional value. In addition, they are low in anti-nutritive factors such as ergot, which have been found in the older varieties.

Triticale should be ground before feeding to improve digestibility. It is also important to ensure the grains are free from ergot (a fungal toxin), to which triticale is more susceptible than other cereals.

Rye

Rye has an energy value intermediate between that of wheat and barley. The protein content is similar to that of barley and oats, but the amino acids are not as well utilized by poultry and livestock. Rye may also contain several toxic anti-nutritional factors, such as ergot.

Like other cereals, this grain should be ground before being used in organic diets. The grinder should be set to break all of the kernels without producing a fine, dusty feed.

SUPPLEMENTARY PROTEIN SOURCES

The major protein sources used in organic diet formulation are soya beans, canola and groundnuts. These crops are grown primarily for their seeds, which produce oils for human consumption and industrial uses. Cottonseed is also used in some diets.

Moderate heating is generally required to inactivate anti-nutritional factors present in oilseeds. Only those meals resulting from mechanical extraction of the oil from the seed are acceptable for use in organic feeding.

As a group, the oilseeds and oilseed meals are high in protein content. Most are low in the amino acid lysine, except soya bean meal. The extent of dehulling affects the protein and fibre contents, whereas the efficiency of oil extraction influences the oil content and thus the energy content of the meal. Oilseed meals are generally low in calcium and high in phosphorus, although a high proportion of the phosphorus is present as phytate.

Canola (Rapeseed) Seed and Meal

Canola is an improved type of rapeseed that is suitable for animal and feeding, the older unimproved types being unsuitable. This crop belongs to the mustard family. It grows best in temperate climates and, as a result, canola is often a good alternative oilseed crop to soya beans in regions not suited for growing soya beans. Some of the canola being grown is from GM-derived seed, therefore caution must be exercised to ensure the use of non-GM canola for organic livestock production.

Canola seed that meets organic standards can be further processed into oil and a high-protein meal, so that the oil and meal are acceptable to the organic industry.

Per kilo, canola seed contains about 400g oil, 230g crude protein and 70g crude fibre. Before mixing into feed, the seed should be ground together with the ground cereal to fracture the seed coat and soak up the released oil. The oil is very unstable and liable to develop rancidity due to its high content of polyunsaturated fatty acids (PUFA).

For organic feed use the oil has to be extracted by mechanical methods such as crushing. The extracted meal is a high quality, high-protein feed ingredient. Canola meal contains about 350–400g crude protein/kg. The lysine content of canola meal is lower and the methionine content is higher than in soya bean meal, otherwise it has a comparable amino acid profile to soya bean meal. However, the amino acids in canola meal are generally less available than in soya bean meal. Owing to its high fibre content (>110g/kg), canola meal contains about 15–25% less energy than soya bean meal.

Cottonseed

Cottonseed is an important oilseed crop, with major producing regions being the USA, China, India, Pakistan, Latin America and Europe. Whole cottonseed is a widely used feed for dairy cattle in North America because of its combination of high fibre, energy (from fat) and protein. Recent data from the USA showed an average concentration of 225 crude protein, 388 acid detergent fibre, 472 neutral detergent fibre, and 178g/kg ether extract in the dry matter. The composition is known to be affected by cultivar.

Cottonseed meal is the second most important protein feedstuff in the USA. As with other oilseeds, only mechanically extracted meal is acceptable in organic feeding. The protein content of cottonseed meal may vary from 360 to 410g/kg, depending on the contents of hulls and residual oil. The fibre content is higher in cottonseed meal than soya bean meal, the energy value being inversely related to the fibre content.

Cattle and sheep are less susceptible than pigs to the presence of an anti-nutritional factor – gossypol – in cottonseed. This is a natural toxin present in the cotton plant that protects it from insect damage. Adult cattle have the ability to detoxify some gossypol in the rumen but this factor limits the levels of cottonseed or cottonseed meal that can be included in calf diets. Ingestion of excess gossypol can be detrimental to fertility in both males and females. This problem does not occur with glandless cultivars of cottonseed.

Groundnuts (Peanuts)

Groundnuts (also known as peanuts) are not included as an approved feedstuff in either the EU or New Zealand lists but should be acceptable for organic diets if grown organically. The reason for omission may be that groundnuts are grown mainly for the human market. This crop is grown extensively in tropical and subtropical regions and is too important to be rejected for use in organic diets. However, this issue should be clarified with the local certifying agency. China and India are the largest producers of groundnuts. Groundnuts not suitable for human consumption are used in the production of groundnut (peanut) oil. The by-product of oil extraction, groundnut meal, is widely used as a protein supplement in poultry and livestock diets in some countries.

Raw groundnuts contain 400–550g oil/kg. Groundnut meal is the ground product of shelled, extracted, groundnuts. Mechanically extracted meal contains about 50–70g oil/kg, and tends to become rancid during storage in hot weather. The protein content of extracted meal ranges from 410 to 500g/kg, hence in North America the meal is usually adjusted to a standard protein level with ground hulls. Groundnut protein is deficient in lysine and is low in methionine and tryptophan. This makes it unsuitable as the sole supplementary protein for most organic diets and it needs to be mixed with other protein sources.

One potential problem with groundnuts is that they are subject to contamination with moulds, such as *Aspergillus flavus*, which produces aflatoxin that is toxic to animals and humans.

Soya Beans and Soya Bean Meal

Soya beans and soya bean meal are generally regarded as the best plant protein sources for animal and poultry feeding. Several bioengineered strains of soya beans are now being grown, therefore organic producers have to be careful to select non-GM products.

Whole soya beans contain 150–210g oil/kg, most of which is removed in the oil extraction process. Only the soya bean meal produced using hydraulic or screw presses (expellers) is acceptable in organic diets. Expeller soya bean meal is favoured for dairy cow feeding since its higher content of rumen by-pass protein results in improved milk production. Consequently much of the expeller soya bean meal is used in the dairy feed industry and may be difficult to obtain by the organic pig or poultry producer.

A process being used in small plants is extrusion, but without removal of the oil, the product being a full-fat meal. Often these plants are operated by co-operatives and should be of interest to organic producers since the product also qualifies for acceptance in organic diets. Another interesting development with soya beans is the introduction of strains suitable for cultivation in cooler climates, for instance the Maritime region of eastern Canada. This development together with the installation of extruder plants allows the crop to be grown and utilized locally, holding the promise for regions that are deficient in protein feedstuffs to become self-sufficient in feed needs. Developments such as

this may help to solve the ongoing problem of an inadequate supply of organic protein feedstuffs in Europe.

Whole soya beans contain 360–370g crude protein/kg, whereas soya bean meal contains 410–500g protein/kg depending on the efficiency of the oil extraction process and the amount of residual hulls present. The oil has a high content of the polyunsaturated fatty acids (PUFA), linoleic and linolenic acids.

Conventional soya bean meal is generally available in two forms, meal with 440g crude protein/kg and meal from dehulled soya beans, which contains 480–500g protein/kg. Because of its low fibre content, the energy content of soya bean meal is higher than in most other oilseed meals. Soya bean meal has an excellent amino acid profile, complementing the limiting amino acids in cereal grains. In addition to the high protein content, the amino acids in soya bean meal are highly digestible in relation to other protein sources of plant origin.

Soya bean meal is generally low in vitamins and minerals, therefore supplementation is necessary.

Natural anti-nutritional factors which interfere with digestion are found in all oilseed proteins. Fortunately they are inactivated when the beans are toasted or heated during processing.

Conventional soya bean meal is one of the most consistent feed ingredients available to feed manufacturers, with the nutrient composition and physical characteristics varying very little between sources. Suppliers of organic soya bean meal need to adopt similar quality control measures to ensure similar consistency in composition.

Full-Fat Soya Beans

Full-fat soya is the complete soya grain (usually after dehulling), containing the original content of oil. It is therefore an excellent source of both energy and protein. The raw bean contains anti-nutritive factors that impair protein digestion but are inactivated by heat treatment (steaming, toasting or extruding). However, if overheated during this process, protein quality will be damaged. Use of full-fat beans is a good way of increasing the energy level of the diet, particularly when they are combined with low-energy ingredients. Also, this is an easier way to blend fat into a diet than by adding liquid fat.

Owing to possible rancidity problems, diets based on full-fat soya beans should be used quickly after mixing and not stored. Otherwise, an approved antioxidant should be added to the diet.

OTHER PROTEIN FEEDSTUFFS

Faba (Field) Beans

This annual legume grows well in regions with mild winters and adequate summer rainfall. Zero-tannin cultivars are reported to yield better than field peas, and they fix atmospheric nitrogen throughout their life cycle. The beans store well for use on-farm.

The beans are often regarded nutritionally as high-protein cereal grains. They contain about 240–300g crude protein/kg, the protein being high in lysine and (like most legume seeds) low in sulphur-containing amino acids. The energy level is similar to that of barley or wheat. The crude fibre content is around 80g/kg, air-dry basis. The oil content of the bean is relatively low (10g/kg dry matter), with a high proportion of the linoleic and linolenic acids. This makes the beans very susceptible to rancidity if stored for more than about a week after grinding. When fresh they are very palatable.

Faba beans contain several anti-nutritional factors, but at low levels

compared with other legume seeds. Improvements in growth rate of poultry and livestock have been recorded when faba beans were roasted or extruded before feeding, possibly due to an improvement in amino acid digestibility or destruction of anti-nutritional factors.

For use in organic diets the beans are usually ground to pass through a 3mm screen.

Field Peas

Field peas are grown primarily for human consumption, but they are now used widely in pig and poultry feeding in several countries. Like field beans, peas are a good cool season alternative crop for regions not suited to growing soya beans, are well suited for early planting on soils that lack water-holding capacity and they mature early. There are green and yellow varieties, which are similar in nutrient content. Some producers grow peas in conjunction with barley, as these two ingredients can be incorporated successfully into organic feeding programmes.

Peas are similar in energy content to high-energy grains such as maize and wheat because of their high starch content. However, since they have a higher protein content (about 230g/kg) than grains they are regarded primarily as a protein source. Pea protein is particularly rich in lysine, but relatively deficient in tryptophan and sulphur-containing amino acids. The oil content is low.

Feed peas, like cereal grains, are low in calcium but contain a slightly higher level of phosphorus. The fibre content is around 55g/kg.

Sunflowers

Sunflower seed is an oilseed crop of considerable potential for organic feeding since sunflowers grow in many parts of the world. It is grown commercially for oil production, leaving the extracted meal available for poultry and livestock feeding. Sunflower seed surplus to processing needs and seed unsuitable for oil production is also available for feed use. On-farm processing of sunflower seed is being done in countries such as Austria.

Per kilo, the seed contains approximately 380g oil, 170g crude protein and 159g crude fibre, and is a good source of energy.

Sunflower meal is produced by extraction of the oil from sunflower seeds. The nutrient composition of the meal varies considerably depending on the quality of the seed, method of extraction and content of hulls. The fibre content of whole (hulled) sunflower meal is around 300g/kg and with complete hull removal is around 120g/kg. Sunflower meal is lower in lysine and higher in sulphur-containing amino acids than soya bean meal. Its energy value is considerably lower than that of soya bean meal. Calcium and phosphorus levels compare favourably with those of other plant protein sources.

Lupins

The low-alkaloid (sweet) types of lupins are becoming increasingly important as a feed ingredient. In Australia, where much of the research on lupins as a feedstuff has been done, the main species used in poultry and pig diets are *Lupinus angustifolius* and *Lupinus luteus*.

The benefits of lupins for the organic producer are that the plant is a nitrogen-fixing legume and, like peas and beans, can be grown and used on-farm with minimal processing. Lupins are also of interest as a forage crop. Another advantage of lupins is that the seed stores well. The shortage of organic protein feedstuffs in Europe has stimulated interest in lupins as an alternative protein source.

One disadvantage of lupins is a high fibre content that makes them unsuitable for young stock unless the seed is dehulled. The energy content depends on the amount of oil present. The seed should be ground before feeding to improve digestibility, but should be used as soon as possible after grinding because the oil is susceptible to rancidity.

The protein content varies with type and has been reported as 272 to 403g/kg (air-dry basis). Like other legumes, the protein is low in methionine and cystine. Dehulling improves the protein content, making it comparable with soya bean meal.

The carbohydrate composition of lupins is different from that in most legumes, with negligible levels of starch and high levels of soluble and insoluble non-starch polysaccharides. This feature influences the utilization of energy in lupins, with more of the carbohydrate being fermented in the large intestine than with other legumes. As a result adult animals, which have more developed digestive systems, are better able to digest lupins than younger animals.

The crude fat content of lupins varies with type (49–130g/kg) and although the fatty acids are mainly unsaturated, the seed contains a high enough level of natural antioxidant to prevent the oil from becoming rancid during storage. This explains why lupins have good storage characteristics.

Lupins are low in most minerals.

Fishmeal

This is one of the best protein sources for organic feeding, since it is a rich source of essential amino acids (particularly lysine), vitamins and minerals. It is a main (often the only) supplementary protein source in Asian countries. For use in organic production, it must be derived from sustainable fish stocks and usually used only in the diet of chicks and young pigs (sometimes in sow lactation diets). Its use is generally limited by cost and availability.

The meal can be produced from the whole fish or from trimmings that are a by-product from human food production. The composition will vary according to the type of fish and the processing. If the meal has been overheated during the drying process, protein quality will be impaired and feed intake may be reduced when fishmeal is included in the diet at high levels.

TUBER ROOTS, THEIR PRODUCTS AND BY-PRODUCTS

Cassava (tapioca, manioc) is grown almost entirely in the tropics and is one of the world's most productive crops, with possible yields of 20 to 30 tonnes of starchy tubers per hectare. Cassava is an approved ingredient in organic diets, although in many countries it will be an imported product not produced regionally. Cassava needs to be heat-treated before being fed to pigs or poultry to reduce the content of cyanogenic glucosides, which release toxic hydrocyanic acid (HCN) in the gut. Cassava meal is an excellent energy source because of its highly digestible carbohydrates (mainly starch). However, its main drawback is the negligible content of protein and micronutrients, requiring a high addition of supplementary protein.

Potatoes

On a worldwide basis this crop is superior to any of the major cereal crops in its yield of dry matter and protein per hectare. Potatoes are now cultivated all over the world except in the humid tropics. In some countries they are grown as a feed crop, in others they are available for animal feeding as cull potatoes or as potatoes surplus to the human market.

Potatoes are used in the commercial production of starch and alcohol, yielding by-products that are potentially useful feedstuffs. The nutritive value of these by-products depends on the industry from which they were derived. Potato protein concentrate is a high quality protein source, whereas potato pulp, the total residue from the starch extraction industry, is a lower quality product for organic feeding because of its higher fibre and lower starch content.

As with most root crops, the major drawback is the relatively low dry matter content (180 to 250g/kg) and low nutrient density. Potatoes are variable in composition, depending on variety, soil type, growing and storage conditions and processing treatment. About 70% of the dry matter is starch and the protein content is similar to that of maize. Potato protein has a high biological value, among the highest of the plant proteins, similar to that of soya beans. Potatoes are essentially a source of energy.

Swedes

There is current interest in root crops such as swedes for organic feeding, as a roughage source and as a crop to allow the animals to express natural foraging behaviour.

Sugar Beet

Sugar beet pulp is a by-product of the sugar industry, comprising the fibre component of the sugar beet root, which may be dried alone (unmolassed sugar beet pulp) or have some of the sugars added back in the form of molasses (molassed sugar beet pulp).

Most sugar beet pulp is now sold after drying and the addition of molasses. The crude fibre content is relatively high (about

200g/kg dry matter) and the crude protein content is low at about 100g/kg dry matter. The fibre is highly digestible and at low inclusion levels can promote good gut health.

Beet molasses is a product of sugar production containing about 700–750g/kg dry matter (mainly sugars) and is difficult to handle as a feed component. The crude protein content is low and is mainly in the form of non-protein nitrogenous compounds. Molasses can be used to improve the palatability of feeds for young stock but only at low levels since it is laxative. Other uses are as a binding agent to reduce dustiness in the feed and as a binding agent in cubes, pellets and the compressed feed blocks that are used as protein, mineral and vitamin supplements for ruminants. Beet molasses is also used as an additive in ensilage.

Cane molasses is obtained as a by-product from the processing of sugar cane into sugar for the human market. It is similar to beet molasses in nutritive content and is used as a feed ingredient in tropical countries where the crop is grown.

ENZYMES

Certain enzymes can be added to organic diets to help the release of nutrients during digestion. Examples are:

1. Phytase, which acts on phytate phosphorus in plant materials, releasing more of the contained phosphorus. As a result, less phosphorus supplementation is required and a lower amount of phosphorus is excreted in manure (possibly 30%).

2. β-Glucanase and xylanase, which can improve the digestibility of carbohydrate, fat and protein in some cereals.

3. α-Galactosidase, which may help the digestion of diets containing high levels of plant protein feedstuffs such as soya bean

meal or lupin seed. These legumes contain oligosaccharides, which cannot be degraded by the gut enzymes and are fermented in the large intestine causing gas production.

4. Addition of α-amylase can improve the digestion of starch and addition of protease has been shown to improve the digestion of protein in some stock.

Producers considering the use of supplementary enzymes should consult a feed supplier and the local certifying agency before doing so.

FORAGES

Pasture is the natural feed for ruminant animals, therefore forages, either grazed or conserved, should comprise a major proportion of the diet of these animals. Forages may be the sole feed provided for low-production stock and when fed fresh or conserved may provide all the nutrients required. Forages are of interest also to organic pig and poultry producers because of the requirement that their diet has to contain some fibre.

In temperate regions, grazed forages are used in late spring, summer and early autumn, while other regions, such as Australia/New Zealand and South America, may support ruminant production on year-round grazing of forages.

The pasture is generally based on grasses (e.g. perennial ryegrass, *Lolium perenne*) with a legume such as white clover (*Trifolium repens*) included in the mix to fix atmospheric nitrogen and improve the nutritive quality of the forage.

In addition to the mix of plants forming the sward, forage quality is also determined by its stage of development and by the soil and climatic conditions. When young and lush, such forage is a feed of high nutritive value and may provide most or all of the requirements of a good dairy ration. Forage species, agronomic conditions, fertilizer practices, maturity at harvest and storage procedures are among the factors that determine the quality of the forage for feeding. The most important grasses worldwide are orchard grass, ryegrass, fescue and timothy. Various species of wheatgrass are used commonly in range conditions for beef cattle in the western USA. Owing to its late maturity, wheatgrass provides a long grazing period when used for pasture. In the early heading stage it is higher in digestible protein and in total digestible nutrients than other wheatgrasses. It produces high yields of hay and can be used for silage. Tall wheatgrass does not exhibit temperature dormancy like many native wheatgrasses, and makes a good recovery after cutting and good autumn growth.

In parts of the USA such as Virginia the primary forage is fescue. However, the presence of a fungal endophyte reduces its suitability for many forage livestock producers because it results in reduced animal productivity.

Legumes are also used as forage crops and increase the quality of pasture, hay or silage. The major species used worldwide are lucerne (alfalfa), clovers and birdsfoot trefoil. Legumes have a lower content of fibre and a higher content of crude protein than grasses. Thus legumes are generally higher in feed value than grasses.

Clovers and other legumes are highly desirable species in pastures and hay meadows (Jennings, 2005) and serve several useful functions. Legumes are able to obtain nitrogen from the atmosphere by means of their symbiotic relationship with Rhizobium bacteria and are, therefore, not dependent on nitrogenous fertilizer. According to this author, under ideal conditions clovers can add up to 240kg N/ha/year to the soil, which can be used by other forage species.

Lucerne

Lucerne (also known as alfalfa) is probably the most important forage legume worldwide. It can be grown over a wide range of soil and climatic conditions and has the highest yield and feeding value of all perennial forage legumes. This perennial herbaceous legume can be used for pasture, hay, silage and green chop, and processed products such as meal, pellets and cubes.

Due to its high nutritional quality, high yields and high adaptability, lucerne is one of the most important legume forages of the world. A major source of protein for livestock, it is a basic component in rations for dairy cattle, beef cattle, horses, sheep, goats and other classes of domestic animals. It is cultivated in more than 80 countries in an area exceeding 35 million ha (Radovic et al., 2009). World production of lucerne was around 436 million tons in 2006.

Genetically modified, glyphosate-resistant lucerne has been developed. Planting started in the USA in 2005 but was regulated in 2007 when a district court found that risks of cross-contamination with non-GM lucerne had not been sufficiently assessed by the USDA. The ban was reversed in June 2010 by the US Supreme Court and planting resumed in January 2011.

Lucerne has a deep root that reaches down to 4m, but can reach 7–9m in well-drained soils. Its stems are erect or decumbent, up to 1m high, and glabrous or hairy in the upper parts. Leaves are trifoliate, with obovate leaflets, 10–45mm long and 3–10mm broad. Inflorescences are oval or rounded racemes bearing 5 to 40 yellow, blue or purple flowers. Fruits are 2–8 seeded curly pods turning from green to brown (Heuzé et al., 2016). There are numerous cultivars of lucerne, selected for specific abilities, such as winter hardiness, drought resistance, tolerance to heavy grazing (such as the Spanish 'Mielga' cultivars) or tolerance to pests and diseases.

Lucerne should be seeded in dense swards if further grazing is planned. It does not tolerate close grazing and some form of rotational grazing is necessary to maintain the persistence and production of plants, with rest intervals that replenish the crown and roots of plants in carbohydrates and nitrogen (Frame, 2005). The duration of rest intervals depends on growth conditions, but 5 to 6 weeks is likely to be necessary. In a continuous grazing system, intensive defoliation can damage the plant crowns. In mixed pastures, stocking rates and grazing intensity should be controlled to prevent the selective overgrazing of lucerne. Some cultivars are better adapted to grazing than others.

Lucerne is usually cultivated for hay, and is frequently used for silage or haylage, dehydrated to make meal or pellets, or used fresh by grazing or cut-and-carry (Radovic et al., 2009; Frame, 2005). In several countries, dehydration plants produce lucerne pellets with standardized protein content targeting specific markets. When managed as a pasture it can be grown as a pure sward or with companion legumes or grasses. Pigment-rich protein concentrates obtained from the aerial part are produced for poultry and pig feed supplementation.

The deep rooting system of lucerne makes it more drought-tolerant than cool season legumes and grasses. Although lucerne does not make maximum growth during summer droughts, it usually provides good summer pastures. During extreme drought, this aspect is even more important since cool season grasses become dormant then.

Grazing can extend the useful life of a stand by 1 year or more for mature lucerne hay fields where some of the stand has been lost or has become weedy. Grazing may also help to rejuvenate some stands by reducing grass and weed competition. Research has shown that lucerne stands with fewer than 30 plants/m^2 may not produce maximum yields of hay.

Alternative Temperate Forages

Researchers in New Zealand have studied alternative temperate forages containing secondary compounds such as condensed tannins for improving sustainable productivity in grazing ruminants (Ramırez-Restrepo and Barry, 2005). Of the forages reviewed, chicory (*Chicorium intybus*) and legumes containing condensed tannins (*Lotus corniculatus*) and sulla (*Hedysarum coronarium*) were recommended. These researchers concluded that the key plant characteristics for improved sustainable productivity were a high ratio of readily fermentable structural carbohydrate and the presence of condensed tannins and certain other secondary compounds.

Conserved Forages

Green chop is very similar to grazed forage except that a machine is used to harvest the crop. Harvesting and storage losses are greatest with hay and silage, but if proper practices are followed, these losses can be minimized.

All of the forage species mentioned above are suitable for organic feeding. Grasses such as Bermuda grass (*Cynodon dactylon*) that are not mentioned specifically in the lists of approved organic feedstuffs are probably acceptable for organic diets, but this should be checked with the local certifying body.

As stated in section 1.6 in Table 2.1 relating to Forages and Roughages, only the following substances are included in this category: lucerne, lucerne meal, clover, clover meal, grass (obtained from forage plants), grass meal, hay, silage, straw of cereals and root vegetables for foraging. They can be conserved by haymaking, silage, etc. As worded, this section applies only to harvested forages and does not apply to grazed forage.

Hay

Haymaking is the traditional method of conserving green crops and is popular with organic producers in Western Europe and other regions. It is made mainly by the sun-drying of grass and other forage crops. After the crop has been cut, its treatment in the field is intended to minimize losses of valuable nutrients caused by the action of plant respiration, microorganisms, oxidation, leaching and by mechanical damage. The aim in haymaking is to reduce the moisture content of the cut crop to a level low enough to inhibit the action of plant and microbial enzymes and allow the hay to be stored satisfactorily for later feeding.

Lucerne is a very important legume, which is grown as a hay crop in many countries. The value of lucerne hay lies in its relatively high content of crude protein, which may be as high as 200g/kg dry matter if it is made from a crop cut in the early bloom stage. Cereals are sometimes cut green and made into hay, usually when the grain is at the 'milky' stage. The nutritive values of cereal hays cut at this stage of growth are similar to those of hays made from mature grass.

The nutritive value of hay is also affected by field losses of nutrients and by changes taking place during storage (which can be reduced by the use of chemical preservatives). Even under good conditions the overall loss of dry matter may be about 20%. Artificially dried forages are higher in nutritive value than hays. However, they are expensive to produce and may be used with non-ruminant livestock as sources of minerals and vitamins.

One point to note is that most weeds are not palatable and in pasture will be avoided by livestock if adequate forage is available. However, most livestock cannot differentiate weeds from beneficial long-stemmed forage in hay, resulting in accidental ingestion and possibly a loss in productivity or death. Thus, an effective weed control programme is recommended.

Hay preservatives may be used to allow hay to be stored at moisture levels that would otherwise result in severe deterioration and moulding. These chemical preservatives include propionic acid, lactic acid-forming bacteria and other biological products. They may be acceptable in organic hay production but this should be checked with the local certifying agency.

The nutritive value of hay is determined by the stage of growth when it is cut and by the plant species of the parent crop. Yields are higher with late cuts but the nutritional value and voluntary intake by cattle are lower. Thus, hay made from early cuts is invariably of higher quality than hay made from mature crops.

During the early phase of drying, enzymes break down or reduce simple sugars and organic acids to carbon dioxide, resulting in an overall loss of dry matter and digestible components. Hence there is a need for drying to be as rapid as possible. Losses of forage dry matter in the field can range from less than 5% to more than 50%, depending on weather conditions and how long it takes the plants to dry. Even under good conditions the overall loss of dry matter may be about 20%. Rainfall can leach plant protein, phosphorus, potassium, carotene and energy components during hay cutting and drying.

The moisture content of a green crop may range from about 650 to 850g/kg, falling as the plant matures. For satisfactory storage the moisture content must be reduced to 150–200g/kg. It is not advisable to wait until the crop is mature and drier before cutting. Moisture content can be measured by taking a sample from the windrow and drying it using a microwave or convection oven. Wet and dry weights can be measured with a scale.

Hay may become mouldy if not dried sufficiently. Mouldy hay is unpalatable to livestock and may be harmful to farm animals and humans because of the presence of mycotoxins. Such hay may also contain actinomycetes, which are responsible for the allergic disease that affects humans known as 'farmer's lung'.

Orchard grass, tall fescue and timothy dry faster and more uniformly than legumes, clovers and ryegrasses. Legume leaf surfaces are more waxy than most grasses, resulting in a slower rate of drying. During the drying process the leaves lose moisture more rapidly than the stems, becoming brittle and readily shattered by handling. If the herbage is bruised or flattened, the drying rates of stems and leaves are more similar. Excessive mechanical handling is liable to cause a loss of leafy material, and since the leaves at the hay stage are higher in digestible nutrients than the stems, the hay produced may be of low nutritive value. Loss of leaves during haymaking is particularly likely to occur with legumes such as lucerne. Baling the crop in the field at a moisture content of 300–400g/kg, and subsequent drying by artificial ventilation, has been shown to reduce mechanical losses considerably.

Artificial drying (known commercially as dehydration) is a very efficient but expensive method of conserving forage crops. It tends to be a commercial process rather than one found on organic farms. In northern Europe, grass and grass–clover mixtures are the crops most commonly dried by this method, whereas in North America lucerne is the primary crop that is dehydrated.

Dehydrated Lucerne

The lucerne dehydration industry developed after the Second World War. Dehydration was found to be a superior method of preservation as it dries and stabilizes the crop while preserving its high protein content, vitamins and overall nutritive value. Also, dehydrated lucerne is a good source of pigments (xanthophylls and beta-carotenes) for poultry feed supplementation.

Dehydration requires pre-wilting and chopping in the field, transportation to the plant and drum-drying (between 250°C and 800°C) down to a moisture content of 100g/kg. After drying, long fibre dehydrated lucerne may be compacted into bales. Lucerne can also be ground to make lucerne meal or ground and passed through a screw die to make pellets. Lucerne is often standardized to a consistent protein content such as 170 or 180g/kg.

As stated above, hays made from legumes are generally higher in protein and minerals than grass hay.

Straw and Chaff

Straw consist of stems and leaves of plants after the removal of the mature seeds by threshing, and are produced from most cereal crops and from some legumes. Chaff consists of the husk or glumes of the seed, which are separated from the grain during threshing. Modern combine harvesters put out straw and chaff together, but older methods of threshing (e.g. hand threshing) yield the two by-products separately. All the straws and related by-products are extremely fibrous. Most have a high content of lignin and all are of low nutritional value.

In rice growing regions rice straw is used as ruminant feed. It is similar to barley straw in nutritional value. In contrast to other straws, the stems are more digestible than the leaves. A by-product similar in composition to straw is sugarcane bagasse, used for ruminant feeding in tropical countries.

Silage

Ensilage is the name given to the process of conserving a crop of high moisture content by controlled fermentation. The product is known as silage. Almost any crop can be preserved as silage, but the most common are grasses, legumes and whole cereals, especially wheat and maize.

There are two major objectives in making hay or grass silage. The first is to remove excess herbage from pasture following its rapid growth in the spring, allowing the land to be grazed subsequently without wastage of surplus grass. The second objective is to conserve the material so that it provides a nutritious feed when grazing is limited or unavailable. Silage production is an alternative to haymaking. Often it is difficult to make hay satisfactorily because of climatic conditions. In order to produce grass hay it is necessary to reduce the moisture content to less than 16% to avoid mould development during storage.

Silage is a good conservation method, even in harsh conditions. Since lucerne has a low carbohydrate content it has to be supplemented with carbon sources, such as ground cereal grains including wheat or barley, and inoculated to start fermentation (Mason, 1998). Lucerne silages can be made using fresh lucerne or pre-wilted lucerne. The crop should be at 50–70% moisture before ensiling to prevent nutrient leaching (Mason, 1998). Making silage with fresh lucerne may cause nutrient losses due to its high water content: it is advisable to limit liquid losses by adding material such as ground cereal grains, sugar beet dehydrated pulp or wheat straw at the bottom of the silo. When the silage is too wet, anaerobic bacteria such as Clostridia may develop and break down the protein. Additives such as organic acids (formic acid, formic acid + formaldehyde, propionic acid) or calcium salts can help to lower pH value and to improve preservation (since a low pH is deleterious to Clostridial development (Mason, 1998). A material that is too dry, not chopped finely enough or insufficiently compacted may result in oxygen being trapped in the silage, resulting in yeast and mould development.

Pre-wilting is the best way to improve forage quality since wilting reduces water content and protein degradation. However, when moisture is above 50%, leaf losses are very important and cause protein loss (Mauriès, 2003).

To ensure a stable fermentation, the ensiled material is stored in a silo or similar container and sealed to maintain anaerobic conditions. The three most important requirements for good silage production are: (i) rapid removal of air; (ii) rapid development of lactic acid, which results in a rapid drop in pH; and (iii) continued exclusion of air from the silage mass during storage and feeding. In practice this is achieved by chopping the crop during harvesting, rapid filling of the silo and adequate consolidation and sealing. After chopping, plant respiration continues for several hours and plant enzymes such as proteases are active until all the air is used up. These enzymes break down the protein in the forage. Rapid removal of air is also important because it prevents the growth of undesirable aerobic bacteria, yeasts and moulds that compete with beneficial bacteria for substrate. If air is not removed quickly, high temperatures and prolonged heating are commonly observed.

Fermentation begins once anaerobic conditions are achieved. Aerobic fungi and bacteria are the dominant microorganisms on fresh herbage, but as anaerobic conditions develop in the silo they are replaced by bacteria able to grow in the absence of oxygen. During the ensilage process these lactic acid bacteria continue to increase, fermenting the water soluble carbohydrates in the crop to organic acids (mainly lactic acid), which reduce the acidity level (pH).

Plant material in the field may range from a pH of about 5 to 6. This decreases to 3.6–4.5 after acid is produced. A rapid reduction in silage pH helps to limit the breakdown of protein in the plant material by inactivating plant proteases. In addition, a rapid decrease in pH inhibits the growth of undesirable anaerobic microorganisms such as enterobacteria and Clostridia. Eventually, continued production of lactic acid and a decrease in pH inhibits growth of all bacteria.

In general, good silage remains stable, without a change in composition or temperature once air is eliminated and the silage has achieved a low pH.

In order to assist in the fermentation process with crops that are difficult to ferment, various silage additives have been used to improve the silage quality. These include live organisms such as lactobacilli, enzymes and propionic acid. Molasses, which is a by-product of the sugar beet and sugarcane industries, was one of the earliest silage additives.

Producers need to check with the local certifying agency to determine whether these additives are acceptable for organic production.

The nutritional value of silage depends upon the species and stage of growth of the harvested crop, and on changes that occurred during the harvesting and ensiling period. Periodic sampling and analysis is therefore recommended.

MICROORGANISMS
Brewer's Yeast

This product (dried *Saccharomyces cerevisiae*) is permitted as a feed ingredient in organic diets, and has been used traditionally in animal diets as a source of nutrients such as vitamins. Inactivated yeast should be used for animal feeding since live yeast may grow in the intestinal tract and compete for nutrients.

Other yeasts have been included in poultry and livestock diets and may be acceptable as ingredients for organic production.

WHICH FEEDSTUFFS ARE BEST SUITED FOR FEEDING THE VARIOUS CLASSES OF ORGANIC POULTRY AND LIVESTOCK?

The above outline tells us which feedstuffs should be used in feeding animals and poultry organically, but it does not indicate which feedstuffs are best suited for each particular class of animal.

Observing which feedstuffs are eaten by poultry and livestock when they are allowed to range outdoors at will gives a good indication of which feedstuffs they prefer. Poultry will peck and scratch the soil to obtain seeds, grains, worms and insects, etc.; pigs will root in the soil to obtain similar food sources but will eat larger items such as tubers; and cattle prefer grass and other forages. These preferences can be used in selecting feedstuffs from the above table for the various classes of farm stock.

Another way of selecting the appropriate feedstuffs is to study the digestive tracts of farmed stock, diagrams of which appear in the following chapters. Cattle and other ruminants have a digestive system that is designed to handle large volumes of bulky feed, distinct from poultry and pigs. Using this information, it is clear that the main feedstuffs for poultry and pigs are grains and protein sources such as soya beans whereas the main feedstuffs for cattle, sheep and goats are forages.

More detailed information on digestibility and utilization of the various feedstuffs described is set out in the species chapters that follow.

A further method of evaluating feedstuffs for use in organic diets is to have the feedstuff analyzed chemically in a laboratory for protein content, etc. The nutritional value of a feed ingredient varies according to the type of animal or bird to which it is fed. For instance, cattle make good use of hay but pigs do not. Nutrition chemists have therefore devised laboratory methods to assess the nutritional composition of feedstuffs, tailored to the type of stock being fed.

Information gained in this way is described below.

LABORATORY ANALYSES OF FEEDSTUFFS

A common procedure, known as Proximate Analysis, is to analyze the contents of major components.

The information gained is as follows.

1. **Moisture** (water): This can be regarded as a component that dilutes the content of nutrients and its measurement provides more accurate information on their actual composition.

2. **Dry matter**: This is the amount of dry material present after the moisture (water) content has been deducted.

3. **Ash**: This provides information on mineral content. Further analyses can provide information on sand contamination and on specific minerals present, such as calcium.

4. **Organic matter**: This is the amount of protein and carbohydrate material present after ash has been deducted from dry matter.

5. **Protein**: This is determined as nitrogen content × 6.25. It is a measure of protein present, based on the assumption that the average nitrogen content is 16g N/100g protein. Some of the nitrogen in most feeds is present as non-protein nitrogen (NPN) and, therefore, the value calculated by multiplying nitrogen content by 6.25 is referred to as crude rather than true protein. True protein is made up of amino acids, which can be measured using specialized techniques.

6. **Non-nitrogenous material**

6.1 **Fibre**: measured as **crude fibre**. Part of this fraction is digestible; therefore more exact methods of fibre analysis have been developed especially for use in cattle, sheep and goat feeding since these species consume forages mainly. The aim is to separate the forage fibre into two fractions: (i) plant cell contents, a highly digestible fraction consisting of sugars, starches, soluble protein, pectin and lipids (fats); and (ii) plant cell wall constituents, a fraction of variable digestibility. The soluble fraction is termed neutral detergent solubles (NDS; cell contents) and the fibrous residue is called neutral detergent fibre (NDF; cell wall constituents). A second method is the acid detergent fibre (ADF) analysis, which breaks down NDF into a soluble fraction containing primarily hemicellulose and some insoluble protein and an insoluble fraction containing cellulose, lignin and bound nitrogen. Lignin has been shown to be a major factor influencing the digestibility of forages.

6.2 **Nitrogen-free extract**: The digestible carbohydrates, i.e. starch and sugars. This is a measure of energy and is calculated as 100 – (% moisture + % crude protein + % crude fat + % crude fibre + % ash).

7. **Fat**: Measured as crude fat (sometimes called oil or ether extract since ether is used in the laboratory extraction process). More detailed analyses can be done to measure individual fatty acids.

Vitamins are not measured directly in the Proximate Analysis system, but can be measured in the fat and water soluble extracts by appropriate methods.

Rapid methods based on techniques such as near infrared reflectance spectroscopy (NIRS) have been introduced to replace chemical methods for routine feed analysis, but bioavailability of nutrients is based mainly on the results of animal studies. A feedstuff or diet can be analysed chemically to provide information on the contents of the components discussed above. Generally this does not provide information on the amount of the nutrient that is biologically available to the animal.

Information on the digestibility of organic feedstuffs in the various species of livestock and poultry, also on the bioavailability of nutrients is outlined in the species sections following.

REFERENCES

Blair, R. (2012). Organic Production and Food Quality: A Down to Earth Analysis. Wiley-Blackwell, Oxford, UK; 282 pp.

European Commission (2007). Council Regulation EC No. 834/2007 on organic production and labelling of organic and repealing regulation (EEC) No. 2092/91. Official Journal of the European Communities, L 189205: 1–23.

Frame, J. (2005). *Medicago sativa L.* In Grassland Index. A searchable catalogue of grass and forage legumes, http://www.fao.org/ag/AGP/AGPC/doc/GBASE/data/pf000346.htm, accessed March 8, 2016.

Heuzé, V., Tran, G., Boval, M., Noblet, J., Renaudeau, D., Lessire, M., Lebas, F., (2016). Alfalfa (Medicago sativa). Feedipedia, a programme by INRA, CIRAD, AFZ and FAO. http://feedipedia.org/node/275, accessed March 18, 2016.

Jennings, J. (2005) Forage Clovers for Arkansas. Publication FSA2117, University of Arkansas Co-operative Extension Service, Fayetteville, Arkansas, 4 pp.

Mason, S., (1998). Alfalfa protein. Alberta Dairy Management, 1A1: 1–2, http://www.agromedia.ca/ADM_Articles/content/alf_prot.pdf, accessed March 5, 2016.

Mauriès, M. (2003). Luzerne: culture, récolte, conservation, utilization. Éditions France Agricole, Group France Agricole, 8, Cité Paradis, 75010, Paris, France.

NZFSA (2011). NZFSA Technical Rules for Organic Production, Version 7. New Zealand Food Safety Authority, Wellington, New Zealand.

Radovic, J., Sokolovic, D. and Markovic, J. (2009). Alfalfa-most important perennial forage legume in animal husbandry. Biotechnology in Animal Husbandry, 25, 465–475.

Ramirez-Restrepo, C.A. and Barry, T.N. (2005). Alternative temperate forages containing secondary compounds for improving sustainable productivity in grazing ruminants. Animal Feed Science and Technology 120: 179–201.

CHAPTER 3

Feeding Organic Pigs

ORGANIC FEEDSTUFFS FOR PIGS

In general the organic regulations state that organic feed must consist of substances that are necessary and essential for maintaining the pigs' health, well-being and vitality and meet their physiological and welfare needs. The regulations include the requirement for roughage and fresh/dried fodder or silage in the daily ration. All feed ingredients used must be certified as being produced, handled and processed in accordance with the standards specified by the certifying body. Pig diets must not include feed medications, growth promoters, lactation promoters, synthetic appetite enhancers, animal by-products, preservation agents, colouring agents and genetically engineered or modified organisms (GMOs) or their products.

The nutritional characteristics, average composition and (where possible) the recommended inclusion rates of the feedstuffs that are considered most likely to be used in organic pig diets are set out below.

Due to a lack of data on feedstuffs that have been grown organically, the nutritional data refer mainly to feedstuffs that have been grown conventionally. Eventually a database of organic feedstuffs composition should be developed.

Jakobsen and Hermansen (2001) summarized the main problems and challenges posed by organic production of pigs in Denmark as follows.

1. To establish the requirements and supply of energy, essential amino acids (EAA), vitamins and minerals under organic farming conditions, using the slow-growing breeds that are preferred over modern hybrids.

   The requirements for vitamins and minerals of pigs under organic farming conditions are not known.
2. To develop feeding concepts in order to improve the resistance to infectious diseases of the gastrointestinal tract.
3. To improve product quality and the economics of production. The quality of organically produced meat is of concern in Denmark, especially in relation to the fatness and the palatability of the meat.

Production results for sows in four large-scale organic pig production units (60–280 sows per unit) in Denmark were inferior to the results obtained in indoor conventional herds due to higher weaning age (49 days) and higher feed consumption.

In addition, the requirement for the inclusion of roughage, fresh or dried fodder or silage in pig diets has to be addressed. To what extent are these ingredients used and do they affect product quality? The high cost of organic grain and protein sources also suggests that producers should explore the maximization of pasture contributions during months when grazing is practical. Should nutrient contributions from pasture be considered when formulating diets? These questions are difficult to answer, given the present state of knowledge. Some Danish research suggested that the vitamins and minerals present in the feed and soil, and synthesis from sunlight, can be used to a higher degree than normally believed. The issue of whether diets for pigs on pasture require supplementation with vitamins and minerals is addressed in a later chapter.

CEREAL GRAINS AND BY-PRODUCTS

The primary sources of energy in pig diets are cereal grains. Also, the processing of cereals for the human market yields by-products that are important as feed ingredients.

In general, cereal grains are low in protein but high in carbohydrate and they are generally palatable and well-digested when passed through a grinder before being included in the diet. Nutrient composition can be quite variable, therefore it is recommended that, where possible, chemical analyses of grain and other main feed products are conducted periodically to monitor their nutrient composition and quality. Analyses for moisture, protein (and possibly lysine) and kernel weight are generally adequate for grains.

Organic cereals tend to have a slightly lower protein content than conventionally grown cereals, due mainly to fertilizer practices and production methods (Blair, 2012).

Yellow maize is the only cereal grain to contain vitamin A, owing to the presence of provitamins (mainly β-carotene). All grains are deficient in vitamin D.

The oil in cereal grains is contained in the germ and varies from less than 20g/kg (dry basis) in wheat to more than 50g/kg in oats. It is high in oleic and linoleic acids, which are unstable after the seed is ground. As a result, rancidity can develop quickly and result in reduced palatability of the feed or feed refusal.

The fibre in grains is contained mainly in the hull (husk) and can be variable, depending on the growing and harvesting conditions. This can affect the starch content of the seed and, as a consequence, the energy value. The hull is quite resistant to digestion and also has a lowering effect on the digestibility of nutrients.

Feed grains can only meet part of the requirement for dietary nutrients, and other feed components are needed to balance the diet completely. Combining grains and other ingredients into a final dietary mixture to meet the pigs' nutritional needs requires information about the nutrient content of each feedstuff and its suitability as a feed ingredient.

Maize, wheat, oats, barley and sorghum are the principal cereals used in pig feeding, as whole grains and/or grain by-products. Generally, maize and wheat are highest in energy value for pigs, with sorghum, barley, oats and rye being lower. Some rye is used in pig feeding. Although it is similar to wheat in composition, it is less palatable and may contain ergot, a toxic fungus. Triticale, a hybrid of rye and wheat, is also being used for pig feeding in some countries.

There do not appear to be any GM (genetically modified) varieties of wheat, sorghum, barley or oats being grown commercially, unlike the situation with maize. In the USA, for instance, substantial quantities of GM maize varieties developed with insect and herbicide resistance are being grown.

These modified varieties are not approved for organic pig production.

Barley has been used traditionally as the principal grain for pig feeding in the western areas of North America, the UK and many countries of Europe because of its better adaptation to climate and the firmer, leaner carcasses produced than with maize-based diets.

Barley is considered a medium-energy grain, lower in energy than maize, and it needs to be ground before use. If no grinding facilities are available, soaking in water for at least 24 hours is an alternative.

Barley contains more fibre than maize and less than oats, but the proportion of hull to kernel is variable, resulting in a variable energy value. The protein content is higher than in maize and can range from about 90 to 160g/kg. The amino acid profile is better than in maize and is closer to that of oats or wheat, and the bioavailability of amino acids is high. Barley contains more phosphorus than other common cereal grains.

This cereal is more suitable in diets for sows and growing-finishing pigs than in diets for young pigs, because of its higher fibre content and lower energy value relative to maize or wheat.

Lower quality barley should be used in dry sow diets. The increased fibre is beneficial in helping to provide gut-fill in these animals on a restricted feeding regimen and helps to develop gut capacity so that the animals are better able to consume the maximum amount of feed during lactation.

Barley can be used in lactation diets, particularly during early lactation when fibre intake may be beneficial in avoiding constipation. If used later in lactation the diets should be supplemented with high-energy ingredients.

Barley-based diets are less popular than maize-based diets for growing-finishing pigs in regions with hot climates because of the higher fibre and resultant higher heat generated from fermentation in the large intestine, which can lead to a decrease in feed intake and a reduction in growth rate.

Hull-less barley varieties have been developed in which the hull separates during threshing. These varieties contain more protein and less fibre than conventional barley, and theoretically should be superior in nutritional value to conventional barley. However, several studies have failed to show improved growth performance over conventional barley in growing-finishing pigs, suggesting that hull-less barley should simply be used in organic pig diets as a substitute for conventional barley on a 1:1 basis.

Barley by-products from brewing can be used in organic pig feeds.

Wheat

This cereal is widely cultivated in temperate countries and in cooler regions of tropical countries.

One concern about wheat is that the energy and protein contents are more variable than in other cereal grains such as maize, sorghum and barley, therefore periodic testing of batches of wheat for nutrient content is recommended. When this is done a wide range of wheat types can be used efficiently in pig diets.

It is commonly used for feeding to pigs when it is surplus to human requirements or is considered not suitable for the human feed market, otherwise it may be too expensive for pig feeding. However, some wheat is grown for feed purposes. By-products of the flour milling industry also are very desirable ingredients for pig diets.

During threshing the husk separates from the grain, leaving a less fibrous product. As a result, wheat is equal to maize in energy value but contains more protein, lysine and tryptophan. Therefore, it can be used as a replacement for maize as a high-energy ingredient and it requires less protein supplementation than maize. It is higher in energy than barley and so is useful in diets for young pigs and lactating sows.

The level used in diets for finishing pigs may have to be curtailed if the animals become too fat and carcass grades are low.

High-moisture wheat preserved by the addition of propionic acid can be used, but since the acid destroys vitamin E and some other vitamins care should be taken to ensure that the diet is supplemented adequately.

When included at high levels in the diet some varieties of wheat can cause digestive upsets and diarrhoea. In such cases the amount in the diet should be reduced.

Wheat is a good ingredient to use in pelleted diets due to its gluten content, which helps to bind the ground feed particles together.

Milling by-products from flour production can also be used in pig feeding. These by-products are usually classified according to their protein and fibre contents and are traded under a variety of names such as pollards, offals, shorts, wheatfeed and midds (wheat middlings). In general, those by-products with low levels of fibre are of higher nutritive value for pigs.

Wheat (Middlings)

The name 'middlings' derives from the fact that this by-product is somewhere in the middle between flour on one hand and bran on the other. This by-product is known as wheatfeed or pollards in Europe and Australia. The composition and quality of middlings vary greatly due to the proportions of fractions included, also the amount of screenings added and the fineness of grind.

Wheat middlings are used commonly in pig diets, as a partial replacement for grain and protein supplement in diets for growing-finishing pigs and sows. This by-product is commonly included in commercial feeds as a source of nutrients and because of its beneficial influence on pellet quality. When middlings (or whole wheat) are included in pelleted feeds, the pellets are more cohesive and there is less breakage and fewer fines.

Oats

Oats are best used in sow gestation diets since they provide a good source of bulk due to the high fibre content. They are less suitable for lactating sows and growing-finishing pigs, where higher energy diets are required. It can be useful to include some oats in the diet of growing-finishing pigs to prevent gastric ulcers. When included at high levels in finishing pig diets, the high content of unsaturated fatty acids in the oil can cause soft carcass fat.

Fine grinding of oats and (where possible) pelleting of the diet improve their utilization by growing-finishing pigs. Coarse grinding of oats can be used with gestation diets.

The inclusion of oats may have to be avoided during hot weather because of a higher heat increment generated by fermentation in the hind-gut, which can result in reduced feed intake and growth rate.

Oats are commonly processed mechanically to remove the hull from the kernel. The result is oat groats. After rolling, the product is known as rolled oat groats. This is a highly palatable, high-energy (3,690kcal DE/kg) feedstuff that is digested well by pigs. The protein content is around 160g/kg. However, this product is too expensive for general use in pig diets and is used mainly in creep feed and pig starter diets.

Maize

Maize is high in starch and in oil and low in fibre, giving it a high energy value. Other grains, except wheat, have a lower energy value than maize. Maize oil has a high proportion of unsaturated fatty acids and is an excellent source of linoleic acid. The use of yellow maize should be restricted in pig diets if it results in carcass fat that is too soft or too yellow.

The protein content of maize is low at about 85g/kg and is not well balanced in amino acid content, lysine and other important amino acids being limiting.

Maize is very low in calcium (about 0.2g/kg). It contains a higher level of phosphorus (2.5–3.0g/kg), but much of the phosphorus is bound in phytate form that is poorly available to pigs. As a result, a high proportion of the phosphorus passes through the gut and is excreted in the manure. The diet can be supplemented with phytase enzyme to improve phosphorus utilization. Another approach is to use one of the newer low phytic acid maize varieties that have about 35% of their phosphorus bound in phytate compared with 70% for conventional maize. These varieties allow the phosphorus to be utilized more effectively by the pig and with less excreted in the manure. As with other improved varieties of maize, producers wishing to use such varieties should check their acceptability with the local organic certifying agency.

The quality of maize is excellent when harvested and stored under good conditions, including proper drying to 100–120g moisture/kg. Fungal toxins can develop in grain that is harvested damp or allowed to become damp during storage, and result in health problems in pigs, especially sows.

Maize is suitable for feeding to all classes of pig. It should be ground medium to medium–fine for use in mash diets and fine for inclusion in pelleted diets. The grain should be mixed into the feed mixture soon after grinding since ground maize is likely to become rancid during storage.

Sorghum

Grain sorghum is generally higher in protein than maize and is similar in energy content. However, one disadvantage of grain sorghum is that it can be more variable in composition because of growing conditions. Protein content usually averages around 89g/kg, but can vary from 70 to 130g/kg. Therefore, a protein analysis prior to formulation of diets is recommended. Lysine is the most limiting amino acid in sorghum, followed by tryptophan and threonine.

Grain sorghum can be used extensively as a cereal grain in pig diets, as a replacement for maize. The hybrid yellow-endosperm varieties are more palatable to pigs than the darker brown sorghums, which possess a higher tannin content. Low-tannin grain sorghum cultivars can be used successfully as the main or only grain source in well-balanced pig growth diets. Proper grinding of grain sorghum is important because of the hard seed coat.

Triticale

This cereal is a hybrid of wheat and rye which grows well on poor soils. It has a similar composition to wheat and is usually slightly higher in essential amino acids. To improve digestibility, it should be ground before feeding. It is important to ensure that this grain is free from ergot, since triticale is more susceptible to this fungal infection than other cereals.

The greatest potential for the use of triticale is in pig operations that grow at least part of their own feed grain supply, using land that is heavily manured. Its greater disease resistance compared with wheat and barley is another advantage. Thus, triticale will be of particular interest to organic pig producers.

Recent research indicates that triticale can replace maize in the diet of growing-finishing pigs provided the content of digestible lysine is maintained.

Rye

Rye has an energy value intermediate to that of wheat and barley, and the protein content is similar to that of barley and oats. Although the amino acid balance is similar to that of barley and wheat, digestibility of the amino acids is 5–10% lower. Rye may also contain several toxic anti-nutritional factors that reduce its nutritional value for the pig, notably ergot.

Whole grains of rye are not completely masticated by pigs and pass through the intestinal tract undigested. Mills should be set to break all the kernels without producing a fine, dusty feed.

SUPPLEMENTARY PROTEIN SOURCES

The major protein sources used in pig feeding are soya beans, canola and groundnuts (peanuts). These crops are grown primarily for their seeds, which produce oils for human consumption and industrial uses. Cottonseed is also used in pig feeding in some regions.

Natural anti-nutritional factors that interfere with digestion are found in all oilseed proteins. Fortunately they are

Component g/kg unless stated	Barley	Wheat	Wheat middlings	Oats	Oat groats	Maize	Sorghum	Triticale	Rye
Dry matter	890	880	890	890	890	890	880	900	880
DE kcal/kg	3110	3365	3075	2770	3690	3525	3380	3320	3270
DE MJ/kg	13.0	14.08	12.87	11.59	15.44	14.75	14.14	13.89	13.68
ME kcal/kg	2910	3210	3025	2710	3465	3420	3340	3180	3060
ME MJ/kg	12.18	13.43	12.66	11.34	14.5	14.31	13.97	13.31	12.8
Crude protein**	113	135	159	115	160	83	92	125	118
Lysine	4.1	3.4	5.7	4	4.5	2.6	2.2	3.9	3.8
Methionine + Cystine	4.8	4.9	5.8	5.8	4.6	3.8	3.4	4.6	3.6
Crude fibre	50	26	73	108	26	26	27	21.6	22
Crude fat	19	20	42	47	60	39	29	18	16
Linoleic acid	8.8	9.3	17.4	19.0	24.0	19.2	13.5	7.1	7.7
Calcium	0.6	0.6	1.2	0.7	0.7	0.3	0.3	0.5	0.6
Phosphorus	3.5	3.7	9.3	3.1	4.5	2.8	2.9	3.3	3.3
Available phosphorus	1.2	1.85	3.8	0.7	1.7	0.4	0.6	1.6	1.1

TABLE 3.1 Average composition of common cereal grains and grain products for organic pig feeding.*

* Air-dry basis. ** N × 6.25

inactivated when the beans are toasted or heated during processing.

Only those meals resulting from mechanical extraction of the oil from the seed are acceptable for organic diets.

As a group, the oilseeds and oilseed meals are high in protein content. Most are low in lysine, except soya beans. The extent of dehulling affects the protein and fibre contents, whereas the efficiency of oil extraction influences the oil content and thus the energy content of the meal. Oilseed meals are generally low in calcium and high in phosphorus, although a high proportion of the phosphorus is present as phytate.

Soya Beans and Soya Bean Meal

Soya beans and soya bean meal are generally regarded as the best plant protein sources for pig feeding. Whole soya beans contain 150–210g oil/kg, much of which is removed in the oil extraction process. Expeller soya bean meal is favoured for dairy cow feeding since its higher content of rumen by-pass protein results in improved milk production. Consequently, much of this product is used in the dairy feed industry and may be difficult to obtain by the organic pig producer.

A process being used in small plants is extrusion, but without removal of the oil, the product being a full-fat meal. Often these plants are operated by co-operatives and should be of interest to organic producers since the product is acceptable for use in organic diets. Another interesting development with soya beans is the introduction of strains suitable for cultivation in cooler climates, for instance the Maritime region of eastern Canada. This development, together with the installation of extruder plants, allows the crop to be grown and utilized locally, holding the promise for regions that

are deficient in protein feedstuffs to become self-sufficient in feed needs. Developments such as this may help to solve the ongoing problem of an inadequate supply of organic protein feedstuffs in Europe.

Whole soya beans contain 360–370g protein/kg, whereas soya bean meal contains 410–500g protein/kg depending on the efficiency of the oil extraction process and the amount of residual hulls present. The oil has a high content of the polyunsaturated fatty acids, linoleic and linolenic acids.

Soya bean meal is generally available in two forms, meal with 440g protein/kg and meal from dehulled soya beans, which contains 480–500g protein/kg. Because of its low fibre content, the energy content of soya bean meal is higher than in most other oilseed meals. Soya bean meal has an excellent amino acid profile, complementing the limiting amino acids in cereal grains. In addition to the high protein content, the amino acids in soya bean meal are highly digestible in relation to other protein sources of plant origin.

Full-Fat Soya Beans

Full fat soya is the complete soya bean after dehulling, and contains the original content of oil. It is therefore an excellent source of both energy and protein.

The raw bean contains anti-nutritive factors that impair protein digestion and can cause gut irritation in young piglets. These factors are inactivated by proper heat treatment (steaming, toasting or extruding).

Use of full-fat beans is a good way of increasing the energy level of the diet, particularly when combined with low-energy ingredients. Also, this is an easier way to blend fat into a diet than by adding liquid fat. Because of its high content of unsaturated fatty acids in the oil, the level of soya beans in the diets of finishing pigs may have to be curtailed to avoid poor grades due to soft backfat.

Full-fat soya beans are particularly useful in lactation diets when feed intake is low and pig pre-weaning survival rate is lower than normal, indicating reduced milk output.

Owing to possible rancidity problems, diets based on full-fat soya beans should be used quickly after mixing and not stored. Otherwise, an approved antioxidant should be added to the diet.

Canola (Rapeseed)

Canola is an improved type of rapeseed that is suitable for pig feeding, the older unimproved types being unsuitable. This crop belongs to the mustard family, and the seed is not as palatable to pigs as soya beans.

Canola seed that meets organic standards can be further processed into oil and a high-protein meal, so that the oil and meal are acceptable to the organic industry.

Per kilo, canola seed contains about 400g oil, 230g protein and 70g crude fibre. The oil is high in polyunsaturated fatty acids (PUFA; oleic, linoleic and linolenic), which makes it valuable for the human food market. It can also be used in animal feed. The oil is, however, highly unstable due to its PUFA content and, like soya bean oil, can result in soft carcass fat unless restricted.

For organic feed use the extraction of oil has to be done by mechanical methods such as crushing. Alternatively, the seed or meal may be processed by extrusion, a process in which the feedstuff is forced through a die under pressure and steam-heated. The result is a high quality, high-protein feed ingredient.

Canola seed should be ground together with the grain portion of the feed mixture, to fracture the seed coat and soak up the oil released during the grinding process.

Canola meal contains about 350–400g

protein per kg. The lysine content of canola meal is lower and the methionine content is higher than in soya bean meal. Otherwise it has a comparable amino acid profile to soya bean meal. However, the amino acids in canola meal are generally less available than in soya bean meal. Because of its high fibre content (>110g/kg), canola meal contains about 15–25% less energy for pigs than soya bean meal.

The response of pigs of all ages to canola meal inclusion in diets is generally favourable. Most of the published research has been conducted using commercial solvent-extracted meal. The results can be used as a guide to the use of expeller canola meal in pig feeding, provided any differences in composition between the two types are taken into account in formulating the diets.

Canola meal can be introduced into starter diets at about 50g/kg once the pigs reach 20kg bodyweight. Later the amount can be increased gradually, to allow all of the supplementary protein in finisher to be supplied as canola meal. Lactating sows and gilts can be fed diets with up to 150g/kg canola meal during lactation and diets with all of the supplementary protein supplied as canola meal during gestation.

Full-Fat Canola (Canola Seed)

A more recent approach with canola is to include the unextracted seed in pig diets, as a convenient way of providing both supplementary protein and energy. Good results have been achieved with this feedstuff, especially with the lower glucosinolate cultivars.

However, there are two potential problems that need to be avoided, as with full-fat soya beans. The maximum nutritive value of full-fat canola is obtained only when the seed is mechanically disrupted and heat-treated to allow glucosinolate destruction and release of the oil. Once ground, the oil in full-fat canola becomes highly susceptible to oxidation, resulting in undesirable odours and flavours. A practical approach to the rancidity problem is to grind just sufficient canola for immediate use.

A conservative recommendation is that the level of full-fat canola in sow diets should be limited to 100g/kg and that the seed be subjected to some form of heat treatment either before or during feed preparation. A similar level is recommended for growing-finishing diets, with a possible reduction during the finishing period if carcasses show a soft fat problem.

Faba (Field) Beans

Faba beans are often regarded nutritionally as high-protein cereal grains. They contain about 240–300g crude protein/kg, the protein being high in lysine and (like most legume seeds) low in methionine and cystine. The energy level is similar to that of barley or wheat. The crude fibre content is around 80g/kg, on an air-dry basis. The oil content of the bean is relatively low (around 10g/kg), with a high proportion of linoleic and linolenic acids. This makes the beans very susceptible to rancidity if stored for more than about a week after grinding. When fresh they are very palatable.

Faba beans contain several anti-nutritional factors, but at low levels compared with other legume seeds. The factor of most concern for pigs is the tannin fraction. Tannins in whole beans are associated with the seed coat, and are lower in white-seeded than in the coloured-seeded varieties. Tannin-free faba beans have lower fibre levels than tannin-containing faba beans, and in research studies have been included successfully in diets for growing-finishing pigs at an inclusion rate up to 300g/kg as a replacement for soya bean meal. On the basis of available data, the suggested

maximal rates of inclusion of faba beans in pig diets are 100g/kg for sows and 200g/kg for grower-finishers, though higher levels have been fed successfully.

Improvements in growth performance have been noted when faba beans are roasted or extruded before feeding, possibly due to an improvement in amino acid digestibility or destruction of anti-nutritional factors.

For use in pig diets the beans are usually ground to pass through a 3mm screen.

Field Peas

Field peas are grown primarily for human consumption, but they are now used widely in pig feeding in several countries. Like field beans, peas are a good cool season alternative crop for regions not suited to growing soya beans. There are green and yellow varieties, which are similar in nutrient content. Some producers grow peas in conjunction with barley, as these two ingredients can be successfully incorporated into a pig feeding programme.

Peas are similar in energy content to high-energy grains such as maize and wheat because of their high starch content. However, since they have a higher crude protein content (about 230g/kg) than grains they are regarded primarily as a protein source. Pea protein is particularly rich in lysine, but relatively deficient in methionine and cystine. The oil content is low.

Feed peas, like cereal grains, are low in calcium but contain a slightly higher level of phosphorus. The fibre content is around 55g/kg.

Results show that peas can be used effectively in pig diets after weaning. They should be ground or pelleted with other feedstuffs when included in pig diets. Research indicates that for pigs weaned at 6 weeks of age, peas can be included in the diet at levels up to 200g/kg.

Peas have been shown to be well accepted as a dietary component by older pigs and that peas can be used as the sole supplementary protein source. One benefit of including peas in pelleted diets is an improvement in pellet quality, due in part to gelatinization of the starch. Inclusion of peas at 100–150g/kg diet has been shown to avoid the need for pellet binders.

The high protein quality and high energy content of peas make them particularly useful in lactating sow diets. Research in Europe showed no differences in wheat/maize (50:50)-based dry sow and nursing sow diets when peas were substituted for all the supplemental protein provided by soya bean meal.

Sunflower Seed

Sunflower is an oilseed crop of considerable potential for organic pig production since it grows in many parts of the world. It is grown commercially for oil production, leaving the extracted meal available for animal feeding. Sunflower seed surplus to processing needs and seed unsuitable for oil production is also available for feed use. On-farm processing of sunflower seed is being done in countries such as Austria.

Per kilo, the seeds contain approximately 380g oil, 170g crude protein and 159g crude fibre and they are a good source of energy for pigs. A general recommendation is that sunflower seeds should not be included in starter, grower or finisher diets at levels above 100–150g/kg. Sunflower seed should be limited to a maximum of 250g/kg in gestating sow diets.

Sunflower meal is produced by extraction of the oil from sunflower seeds. The nutrient composition of the meal varies considerably, depending on the quality of the seed, method of extraction and content of hulls. Its energy value is considerably lower than that of soya bean meal. The crude fibre content of whole (hulled) sunflower meal is around 300g/kg and with complete hull removal is around 120g/kg. Sunflower meal is lower in lysine and higher in methionine and cystine than soya bean meal.

Because of its high fibre content and low energy value it should be used in limited quantities in pig diets. Sunflower meal containing high levels of oil can produce soft backfat because of the unsaturated fatty acids in the oil. Most studies report decreased growth performance in weaners and growing-finishing pigs when sunflower meal completely replaces soya bean meal, due mainly to its low lysine content. As a result, sunflower meal can be used only to form part of the supplementary protein needed in the diet for growing-finishing pigs.

Lupins

Seed from the low-alkaloid (sweet) types of lupins are becoming of increasing importance as a feed ingredient. In Australia, where much of the research on lupins as a feedstuff for pigs has been done, the main species of lupins used in pig diets are *Lupinus angustifolius* and *L. luteus*. The research indicated that *L. albus* is not currently recommended for use in conventional pig diets because it causes a reduction in growth rate.

One disadvantage of lupin seed is a high fibre content that makes it unsuitable for young pigs unless dehulled. The energy content depends on the amount of oil present. The seed should be ground before feeding to improve digestibility, but should be used as soon as possible after grinding because the oil can go rancid due to its content of polyunsaturated fatty acids.

The crude protein content varies with type and has been reported as 272 to 403g/

kg (air-dry basis). Like other legumes, the protein is low in methionine and cystine. Dehulling improves the protein content, making it comparable with soya bean meal.

Gas production is higher with diets containing lupin seed, some research indicating that when lupin seed is included in sow diets at levels above 200g/kg diet the excess gas production may compromise sow health.

Despite this feature lupin seed is an excellent energy source for pigs. The sow has a high capacity for hindgut fermentation of lupin kernel and hulls.

The crude fat content of lupins varies with type (49–130g/kg) and, although the fatty acids are mainly unsaturated, the seed contains a high enough level of natural antioxidant to prevent the oil from becoming rancid during storage. This explains why lupins have good storage characteristics.

Good results have been reported when ground lupin seed from *L. angustifolius* was included in pig diets. Australian researchers recommended maximum dietary inclusion levels of 200–250g/kg for growing pigs of 20 to 50kg live weight, 300–350g/kg for finishing pigs of 50 to 100kg live weight, and 200g/kg for dry sow and lactation diets (Blair, 2007). These results are applicable to low-alkaloid types of lupin. More conservative levels should be used when the alkaloid content is uncertain since it is known that the lupin alkaloids will depress productivity and the health of pigs.

Groundnuts

Groundnuts (also known as peanuts) are not included as an approved feedstuff in either the EU or New Zealand lists but should be acceptable for organic pig diets if grown organically. Groundnuts not suitable for human consumption are used in the production of groundnut (peanut) oil. The byproduct of oil extraction, groundnut meal, is widely used as a protein supplement in pig diets in some countries.

One potential problem with groundnuts is that they are subject to contamination with moulds, such as *Aspergillus flavus*, which produces aflatoxin that is toxic to animals and humans. Periodic testing for mould contamination is recommended.

Raw groundnuts contain 400–550g oil/kg. Groundnut meal is the ground product of shelled groundnuts. Mechanically extracted meal contains about 50–70g oil/kg, and tends to become rancid during storage in hot weather. The crude protein content of extracted meal ranges from 410 to 500g/kg. Groundnut protein is deficient in lysine and is low in methionine and tryptophan. This makes it unsuitable as the sole supplementary protein for pig diets and it requires mixing with other protein sources.

Owing to its low lysine content, using groundnut meal in combination with ingredients high in lysine appears to be the most effective way to use this protein supplement in pig diets. Groundnut meal is a good protein source for pig diets in countries such as Nigeria that are able to grow this crop, and has been shown to be superior to cottonseed cake for use in maize-based diets for growing pigs.

Whole groundnuts surplus to human needs or of a grade unacceptable for human use can be used as feedstuffs for pigs. Research suggests that a level of 50g/kg diet is optimal for weanling pigs and 50–100g/kg for growing pigs. Their inclusion may have to be limited to less than 100g/kg in diets for market pigs to prevent soft backfat.

Fishmeal

This is one of the best protein sources for pigs, since it is a rich source of essential amino acids (particularly lysine), vitamins and minerals. It is a main (often the only)

Component g/kg unless stated	Soya beans	Soya bean meal	Canola seed	Canola meal	Groundnut meal	Faba beans	Field peas	Sunflower meal dehulled	Lupin seed
Dry matter	900	900	940	940	930	870	890	930	890
DE kcal/kg	4140	3610	4330	3011	3895	3245	3435	3113	3630
DE MJ/kg	17.32	15.1	18.12	12.6	16.3	13.58	14.37	13.02	15.19
ME kcal/kg	3960	2972	4157	2880	3560	3045	3210	2737	3380
ME MJ/kg	16.57	12.43	17.4	12.05	14.9	12.74	13.43	11.45	14.14
Crude protein**	352	429	242	352	432	254	22.8	414	349
Lysine	22.2	27.9	14.4	27	14.8	16.2	15	16.1	15.4
Methionine + Cystine	10.8	12.1	8.6	19	11	5.2	5.2	16.3	7.8
Crude fibre	43	59	74	120	69	73	61	122	110
Crude fat	180	48	397	110	65	14	12	80	97.5
Linoleic acid	104.6	27.9	110	12.8	17.3	5.8	4.7	19	35.6
Calcium	2.5	2.6	3.9	7.6	1.7	1.1	1.1	3.9	2.2
Phosphorus	5.9	6.1	6.4	11.5	5.9	4.8	3.9	10.6	5.1
Available phosphorus	2.3	2.3	1.9	3.4	0.7	1.9	1	2.7	1

TABLE 3.2 Average composition of common protein feedstuffs for organic pig feeding.*

* As-fed basis. ** N × 6.25

supplementary protein source in Asian countries.

For use in organic production, it must be derived from sustainable fish stocks and usually is used only in the diet of young pigs (sometimes in lactation diets). Its use is generally limited by cost and availability.

It is recommended that the amount of fishmeal in pig diets should not exceed 60–70g/kg and should be avoided entirely in finisher diets because of its potential to cause a fishy flavour in pork.

TUBER ROOTS, THEIR PRODUCTS AND BY-PRODUCTS

Cassava

Cassava meal is an excellent energy source because of its content of highly digestible carbohydrates (mainly starch). However, its main drawback is the negligible content of protein and micronutrients.

Cassava needs to be heat-treated before being fed to pigs to reduce the content of cyanogenic glucosides, which release toxic hydrocyanic acid (HCN) in the gut. It is recommended that pig diets contain no more than 100mg HCN equivalent/kg.

Correctly processed cassava can replace maize in growing-finishing pig diets, up to a maximum inclusion level of 400g/kg diet. Dustiness of cassava-based diets can be reduced by adding molasses or oils and by pelleting. Limited studies suggest that cassava can be incorporated successfully in gestation and lactation diets for sows.

Potatoes

As with most root crops, the major drawback is the high moisture and relatively low dry matter content (180 to 250g/kg) and low nutrient density. About 70% of the dry matter is starch and the protein content is similar to that of maize. Thus potatoes are essentially a source of energy. Potato protein has a high biological value, among the highest of the plant proteins, similar to that of soya beans.

Raw potatoes are unpalatable to pigs and are not well digested because the starch and protein are resistant to breakdown by gut enzymes. Therefore raw potatoes should be boiled or steamed before being fed. Ensiling is ineffective in dealing with the factors associated with low palatability and low digestibility in raw potatoes.

For best results potatoes should be boiled for 30–40min, steamed at 100°C for 20–30min or simmered for 1 hour, to ensure thorough cooking through to the centre, and then cooled rapidly. Prolonged heating, or slow cooling after heating, results in damage to the protein and a reduction in its digestibility. Green potatoes should be avoided for organic feeding since they contain solanin, which is poisonous.

Research has shown that cooked potato is approximately equal to maize in energy and digestible protein contents and is palatable to pigs of all ages.

Growth rates have been shown to be similar when cooked potato replaced ground maize in grower-finisher diets. One benefit of potato feeding was a firm backfat. Another finding was that pellet quality of the diets was improved, diets containing potato flakes or flour being firmer than those containing cereals. To avoid gestating sows becoming too fat it is recommended that they be restricted to a daily ration of 4–6kg of cooked potato plus 1kg of a supplementary concentrate. Lactating sows can receive a daily ration of 5–7kg plus a higher amount of supplement.

Potato By-products

Potato by-products include potato meal, potato flakes, potato slices and potato pulp. These products are very variable in their nutritive value depending on the processing method, therefore should be purchased on the basis of a guaranteed analysis.

Dehydrated cooked potato flakes or flour are very palatable and can be used as a cereal replacement. Owing to the cost, they are generally limited to diets for young pigs or lactating sows.

Potato Protein Concentrate

Potato protein concentrate is a high quality product, widely used in the human food industry because of its high digestibility and high biological value of the protein. It is an excellent protein and amino acid source for use in all pig diets. However, its high cost restricts it to diets for young pigs where it can replace milk and fish protein sources at up to 150g/kg.

Swedes

There is current interest in root crops such as swedes for organic pig feeding, as a roughage source and as a crop to allow the animals to express natural foraging behaviour.

Research studies indicate that swedes can be used as a replacement for barley at a rate of about 1.5 units of swede dry matter replacing 1 unit of barley dry matter. Pigs given the diets containing swedes were able to consume daily amounts of dry matter equivalent to that of pigs receiving a control diet, but that continuous access to the feed trough was necessary. This finding indicates that when bulky feeds are used, sufficient feed trough space must be provided to allow all animals in the group to feed at all times.

Sows appear to like swedes and when they have access to this crop spend less time rooting in pastures. Research findings suggest that the daily concentrate ration could be reduced by 0.5kg in sows provided with access to swedes.

Sugar Beet

Sugar beet pulp is a by-product of the sugar industry, comprising the fibrous component of the sugar beet root, which may be dried alone (unmolassed sugar beet pulp) or have some of the sugars added back in the form of molasses (molassed sugar beet pulp). The fibre is highly digestible and at low inclusion levels can promote good gut health. However, it readily absorbs water and swells, reducing voluntary food intake at high levels. This can be a problem in young piglets, but can reduce hunger in limit-fed dry sows and assist in the improvement of carcass grading in finishing pigs.

Food Processing By-products

Although not specified as acceptable ingredients of organic diets for pigs, food by-products from the processing of organic foods for the human market have potential as a valuable feed resource. One assessment of this potential was carried out by Wlcek and Zollitsch (2004), who studied the possibility of recycling organic food by-products by feeding them to organically raised pigs.

Sector-specific questionnaires were sent to 321 processors of organic foods in Austria and proximate analysis, amino acid and mineral analyses were performed on each by-product. It was found that annually 2,400 tonnes of wheat bran, 990 tonnes of rye bran and 1,300 tonnes of residues from the processing of cereal grains were already being fed to livestock. Some 510 tonnes of stale bread was disposed of, but could be used as a highly nutritive feedstuff for pigs.

Substantial quantities other energy-rich by-products were identified, namely about 11,000 tonnes (2,000 tonnes on a dry matter basis) of feed-grade potatoes that were currently being composted, resulting in a waste of 27,000 GJ of metabolizable energy. These potatoes could be better utilized as a dietary energy source for approximately 12,300 pigs. Additionally, about 12,900 tonnes of whey from organically produced milk was being discarded, which could be used to feed roughly 14,000 pigs. High-protein by-products were less available, about 80 tonnes and 63 tonnes of expeller processed pumpkin seed and sunflower seed, respectively, were being produced annually from organically grown oilseeds. Only small quantities of okara (a by-product of the production of tofu from soya beans) and buttermilk were available.

On the assumption that the total amounts of all available by-products were used in Austria as feedstuffs for pigs, the authors calculated that the supply of protein, lysine and energy could be as high as 42%, 31% and 37%, respectively, of the theoretical requirements. They calculated also that utilisation of the organic feed grade potatoes could allow the raising of about 19% of all organic slaughter pigs and save about 1,700 tonnes of other feedstuffs (mainly cereal grains). The saving of other feedstuffs was seen as being particularly important because of the scarcity of organic feedstuffs and likely future restrictions on the use of conventional feedstuffs in organic diets.

Other countries probably have a similar potential for using organic food wastes in organic pig production.

FORAGES AND ROUGHAGES

Forages are of interest to organic pig producers because of the requirement that the diet of organic pigs has to contain roughage, fresh (grazing) or dried fodder or silage.

Approved feedstuffs in this section include lucerne, lucerne meal, clover, clover meal, grass (obtained from forage plants), grass meal, hay, silage, straw of cereals and root vegetables for foraging. They can be conserved by haymaking, ensilage, etc.

Lucerne (also known as Alfalfa)

This is a perennial herbaceous legume. Due to its high nutritional quality, high yields and high adaptability, lucerne is one of the most important legume forages of the world. It is used mainly for ruminant animals such as dairy cattle, beef cattle, horses, sheep and goats but is also used in pig feed, both in processed form and as a forage.

Lucerne does not tolerate close grazing, and some form of rotational grazing is necessary to maintain the persistence and production of plants. When managed as a pasture it can be grown as a pure sward or with companion legumes or grasses.

It is usually cultivated for hay, and is frequently used for silage or haylage, dehydrated to make meal or pellets, or used fresh by grazing or cut-and-feed. In several countries, dehydration plants produce lucerne pellets. Sun-cured lucerne has a protein content higher than most cereal grains. It is high in calcium, low in phosphorus, and is a good source of other minerals and of most vitamins.

Hay made from early cuts is generally of higher quality than hay made from mature crops.

Silage is a good conservation method, with the crop at 50–70% moisture before ensiling to prevent nutrient leaching. Since lucerne has a low carbohydrate content it has to be supplemented with carbon sources, e.g. ground cereal grains such as wheat or barley.

Component g/kg unless stated	Cassava	Molasses Beet	Molasses Cane	Potatoes cooked	Potato protein concentrate	Sugar beet pulp dried	Swedes
Dry matter	880	770	710	222	910	910	103.4
DE kcal/kg	3385	2329	2469	956	4140	2865	406
DE MJ/kg	14.16	9.71	10.32	4.0	17.32	12	1.7
ME kcal/kg	3330	2227	2343	918	3880	2495	390
ME MJ/kg	13.93	9.32	9.8	3.84	16.23	10.44	1.63
Crude protein**	33	60	29	24.5	755	86	12
Lysine	0.7	0.1	NA	1.43	64	5.2	0.46
Methionine + Cystine	0.4	0.1	NA	0.7	32	1.3	0.27
Crude fibre	NA	0	0	7.5	5.5	182	12
Crude fat	5	0	0	1.0	20	8	NA
Linoleic acid	NA	0	0	0.01	NA	NA	0.35
Calcium	2.2	2	8.2	0.14	1.7	7.0	0.55
Phosphorus	1.5	0.3	0.8	0.7	2.0	1.0	0.68
Available phosphorus	0.7	0.2	NA	0.6	2.0	0.5	NA

TABLE 3.3 Average composition of common tuber roots and forage by-products, etc., for organic pig feeding.*

* As-fed basis. ** N × 6.25

Dehydrated Lucerne (Dehy)

Dehydration is a good way to preserve lucerne for later use, though the process is usually done off-farm. This preserves the maximum content of nutrients. The dehydrated products can subsequently be ground into a meal or made into pellets.

The protein content of dehydrated lucerne is about 170g/kg, making it a potential protein supplement for pig diets. However, the fibre content is high at around 250 to 300g/kg and makes the protein harder to digest by pigs (around 60%). This limits the amount that can be included in pig diets.

Suggested maximum inclusion rates of feedstuffs in organic pig diets are shown in Appendix Table 3.16.

FEED MATERIALS OF MINERAL ORIGIN

The approved products to supply the mineral requirements of pigs include the following:

Sodium products: Unrefined sea salt, coarse rock salt, sodium sulphate, sodium carbonate, sodium bicarbonate and sodium chloride.

Calcium products: Lithotamnion and maerl shells of aquatic animals (including cuttlefish bones), calcium carbonate, calcium lactate, calcium gluconate.

Phosphorus products: Bone dicalcium phosphate precipitate, defluorinated dicalcium phosphate and defluorinated monocalcium phosphate.

The approved products to supply the trace element requirements include the following:

Iron products: Ferrous carbonate, ferrous sulphate monohydrate and ferric oxide.

Iodine products: Calcium iodate, anhydrous calcium iodate hexahydrate and potassium iodide.

Copper products: Copper oxide, basic copper carbonate monohydrate and copper sulphate pentahydrate.

Zinc products: Zinc carbonate, zinc oxide, zinc sulphate monohydrate and/or heptahydrate.

Selenium products: Sodium selenate and sodium selenite.

VITAMINS

The vitamins approved for use in organic feed have to be preferably derived from ingredients occurring naturally in feeds, or synthetic vitamins identical to natural vitamins (related EU regulation, Directive 70/524/EEC). When the organic feed or organic pork product is to be exported to the USA, the vitamins and trace minerals used have to be FDA-approved.

Enzymes authorized under EU Directive 70/524/EEC and regulated regulations.

Microorganisms authorized under EU Directive 70/524/EEC.

ENZYMES

Certain enzymes can be added to organic diets, to help the release of nutrients during digestion (related EU regulation, Directive 70/524/EEC). Examples are:

1. Phytase, which acts on phytate phosphorus in plant materials, releasing more of the contained phosphorus. As a result less phosphorus supplementation is required and a lower amount of phosphorus is excreted in manure (possibly 30%).
2. β-Glucanase and xylanase, which can improve the digestibility of carbohydrate, fat and protein in some cereals.
3. α-Galactosidase, which may be help the digestion of diets containing high levels of plant protein feedstuffs such as soya bean meal or lupin seed. These legumes contain oligosaccharides, which cannot be degraded by the gut enzymes and are fermented in the large intestine causing gas production.
4. Addition of α-amylase can improve the digestion of starch and addition of protease has been shown to improve the digestion of protein. These applications are of more relevance to very young pigs and are less relevant to organic production since the animals are older when weaned.

Producers considering the use of supplementary enzymes should consult with a feed supplier and the local certifying agency before doing so.

MICROORGANISMS

Certain microrganisms are approved for use in organic pig feed (related EU regulation, Directive 70/524/EEC).

Brewer's Yeast

This product (dried *Saccharomyces cerevisiae*) is permitted as a feed ingredient in organic diets, and has been used traditionally in animal diets as a source of nutrients.

Inactivated yeast should be used for animal feeding since live yeast may grow in the intestinal tract and compete for nutrients.

Other yeasts have been included in pig diets and may be acceptable as ingredients for organic production.

OTHER ADDITIVES APPROVED FOR ORGANIC PIG FEEDING

The regulations permit the addition of certain additives for use in ensilage, as processing aids, pellet binders, etc. These are detailed in Chapter 2, Table 2.1.

DIGESTION AND ABSORPTION OF NUTRIENTS

An understanding of how well pigs are able to digest the feed sources available and how well the products of digestion meet the needs for maintenance, growth and production is necessary so that suitable feed mixtures can be formulated.

Like all other animals, pigs require five components in their diet as a source of nutrients: energy, protein, minerals, vitamins and water. A nutrient shortage or imbalance in relation to other nutrients will adversely affect growth, health and production. Pigs need a well-balanced and easily digested diet for optimal reproduction and meat production. They are very sensitive to dietary quality because they grow quickly in relation to their body weight and make relatively little use of fibrous, bulky feeds such as lucerne hay or pasture. The digestive system in pigs is quite like that in humans, and because they have a simple stomach compartment (unlike that in cattle and sheep) they are termed non-ruminants (Figure 3.1).

A summary outline of digestion and absorption in the pig follows. This provides a basic understanding of how the feed is digested and the nutrients absorbed. Readers interested in a more detailed explanation of this topic should consult a recent text on pig nutrition.

Digestion is the preparation of feed for absorption, i.e. reduction of feed particles in size and solubility by mechanical and chemical means. Mechanical breakdown of

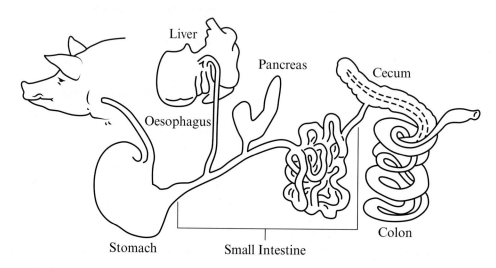

FIGURE 3.1 Digestive system of pigs.

feedstuffs is performed by chewing in the mouth and contractions of the muscles of the gastrointestinal walls. Chemical breakdown is achieved by enzymes secreted in digestive juices and by gut microflora. The digestive process reduces feed particles to a size and solubility that allows for absorption of digested nutrients through the gut wall into blood and lymph.

Mouth

Digestion begins in the mouth. Here feed is chewed into smaller pieces, which increases its surface area. This aids subsequent chemical reactions with various digestive juices and enzymes. Saliva produced in the mouth by the salivary glands moistens the dry feed so that it is easier to swallow. At this point the feed is tasted and, if accepted, swallowed. The saliva is slightly acidic. Pigs produce about 15 litres of saliva daily. Pigs are the only farm animals whose saliva contains the enzyme ptyalin, which has amylase activity, but it is doubtful whether much starch is digested in the mouth since the feed passes quickly to the stomach by means of a series of muscular contractions (peristalsis) in the oesophagus.

Stomach

The stomach capacity of a 90kg pig is 6–8 litres. A churning action here further softens and separates feed particles, exposing them to gastric juices secreted by the stomach. These juices contain several enzymes, principally pepsin, which act to break down protein. Pepsin can function only in an acid medium (pH less than 3.5), acidity being provided by hydrochloric acid produced by the stomach. Hydrochloric acid dissolves minerals ingested with the feed such as calcium salts and inactivates pathogenic bacteria present in the feed. Mucus is

also released by the stomach to protect the stomach wall from acid damage. The small amount of lipase present in gastric juices initiates the digestion of fat in the stomach. In nursing pigs, the gastric juices also include the enzyme rennin, which breaks down the protein in milk. Thorough mixing of the feed with acidic gastric juices takes place only in the lower region of the stomach. The contents of the upper stomach remain at an alkaline pH. This enables digestion of starch in the slightly alkaline medium of the stomach by salivary amylase. One or two hours after eating, partially digested feed in a semi-fluid form known as chyme moves from the stomach into the small intestine.

Small Intestine

The small intestine is a long tube-like structure connecting the stomach to the large intestine. This is where digestion is completed and absorption of nutrients takes place. Its capacity in the pig is about 9–10 litres. Absorption includes various processes that allow the end products of digestion to pass through the membranes of the intestine into the bloodstream for distribution throughout the body. Some absorption also takes place in the stomach.

Chyme is mixed with other fluids in the small intestine, the first part of which is known as the duodenum. Duodenal glands produce an alkaline secretion that acts as a lubricant and also protects the duodenal wall against hydrochloric acid entering from the stomach. The pancreas (which is attached to the small intestine) secretes fluid containing bicarbonate and several enzymes (amylase, trypsin, chymotrypsin and lipase) that act on carbohydrates, proteins and fats. The duodenal wall also secretes enzymes that continue the breakdown process on sugars, protein fragments and fat particles. Bile synthesized by the liver passes into the duodenum via the bile duct. It contains bile salts that provide an alkaline pH in the small intestine and fulfil an important function in digesting and absorbing fats. As a result of these activities the ingested carbohydrates, protein and fats are broken down into small molecules. Muscles in the wall of the intestine regularly contract and relax, mixing the chyme and moving it towards the large intestine.

Newborn pigs for a short time after birth possess the ability to absorb large molecules in a manner similar to that by which an amoeba surrounds its food source (pinocytosis). This is important in that it allows newborn pigs to receive immunoglobulin from colostrum, which provides some immunity against diseases in the environment. Another feature of the young pig is its great ability to digest fat from sow's milk, which is high in this component.

Jejunum and Ileum

Absorption also takes place in the second section of the small intestine, known as the jejunum, and in the third section, known as the ileum. Digestion and absorption are complete by the time the ingesta have reached the terminal end of the ileum. This area is therefore of interest to researchers studying nutrient bioavailability (relative absorption of a nutrient from the diet) since a comparison of dietary and ileal concentrations of a nutrient provides information on its removal from the gut during digestion and absorption. Minerals and vitamins are not changed by enzymatic action. They dissolve in various digestive fluids and water, and are then absorbed.

Once the nutrients enter the bloodstream, they are transported to various parts of the body for vital body functions. Nutrients are used to maintain essential functions such as breathing, circulation of blood and muscle

movement, replacement of worn out cells (maintenance), growth, reproduction and secretion of milk (production).

The ingesta, consisting of undigested feed components, intestinal fluids and cellular material from the abraded wall of the intestine, then passes to the next section of the intestine, the large intestine.

Large Intestine

The large intestine (lower gut) consists of two parts: a sac-like structure called the caecum and the last section, called the colon. The colon is attached to the rectum. The caecum is small, with a capacity of about 1.5–2 litres. No enzymes are released into this part of the gut. Some microbial breakdown of fibre and undigested material occurs, but absorption is limited. Thus fibrous feeds, such as lucerne, have limited feed value.

Remaining nutrients, dissolved in water, are absorbed in the lower part of the colon (about 9 litres capacity). The nutritional significance of certain water soluble vitamins and proteins synthesized in the large intestine is doubtful because of limited absorption in this part of the gut. The large intestine absorbs much of the water from the intestinal contents into the body, leaving the undigested material, which is formed into the faeces and later expelled through the anus.

The entire process of digestion requires about 24–36 hours.

DIGESTIBILITY

Only a fraction of each nutrient taken into the digestive system is absorbed. This fraction can be measured as the digestibility coefficient. It is determined through animal digestibility experiments. Researchers measure both the amount of nutrient present in the feed and the amount of nutrient present in the faeces (or more exactly in the ileum). The difference between the two, commonly expressed as a percentage or in relation to 1 (1 indicating complete digestion), is the amount of the nutrient digested by the pig. Each feedstuff has its own unique set of digestibility coefficients for all nutrients present. The digestibility of a feedstuff or a complete feed can also be measured. Digestibility measured in this way is known as apparent digestibility since the faeces and ileal digesta contain substances originating in the fluids and mucin secreted by the gut and associated organs, as well as cellular material abraded from the gut wall as the digesta pass. Correction for these endogenous losses allows true digestibility to be measured.

Generally, the digestibility values listed in feed tables refer to apparent digestibility unless stated otherwise.

NUTRIENT REQUIREMENTS

Energy

Energy is produced when the feed is digested in the gut. The energy is then either released as heat or is trapped chemically and absorbed into the body for metabolic purposes. It can be derived from protein, fat or carbohydrate in the diet. In general, cereals and fats provide most of the energy in the diet. Energy in excess of requirement is converted to fat and stored in the body. The provision of energy accounts for the greatest percentage of feed costs.

The total energy (gross energy: GE) of a feedstuff can be measured in a laboratory by burning it under controlled conditions and measuring the energy produced in the form of heat. Digestion is never complete under practical situations, therefore measurement

of GE does not provide accurate information on the amount of energy useful to the animal. A more precise measurement of energy is digestible energy (DE), which takes into account the energy lost during incomplete digestion and excreted in the faeces.

Much research has been conducted on this topic and there is now a large database of DE values for feedstuffs, which is used widely by nutritionists in formulating pig diets.

The chemical components of feedstuffs have a large influence on DE values, with increased fat giving higher values and increased fibre and ash giving lower values. Fat provides about 2.25 times the energy provided by carbohydrates or protein. These relationships have been quantified by researchers to allow DE values to be predicted from chemical composition when no determined DE values are available.

More accurate measures of useful energy contained in feedstuffs are metabolizable energy (ME: which takes into account energy lost in the urine) and net energy (NE: which in addition takes into account the energy lost as heat produced during digestion).

However, these measures are more difficult to obtain and in the case of ME may be calculated from DE rather than measured directly. As a result, DE is the most common energy measure used in pig nutrition, at least in North America. The ME system is widely used in Europe and to some extent in North America. Where ME is used it is often calculated from DE, using a factor of 0.96 (ME=DE × 0.96). Feed composition tables may quote ME values, but many of these have been calculated from DE using this or a similar factor.

The requirements set out in this publication and taken mainly from the report on the Nutrient Requirements of Swine (NAS–NRC, 1998) were compiled from results obtained with pigs raised conventionally, not organically.

The NAS–NRC report based their energy requirement on DE, expressed as kilocalories (kcal) or megacalories (Mcal) per kg feed. This energy system is used widely in North America and in many other countries. Energy units used in some countries are based on joules (J), either kilojoules (kJ) or megajoules (MJ). A conversion factor can be used to convert calories to joules, i.e. 1Mcal = 4.184MJ, 1MJ = 0.239Mcal and 1 MJ = 239kcal. Therefore the tables of feedstuff composition in this publication show DE and ME values expressed as MJ or kJ as well as kcal per kg.

Some countries in Europe such as Denmark and the Netherlands use feeding systems based on NE. These systems are more precise than systems based on DE or ME but they need the feed composition and requirements to be based on NE, the data for which are not yet widely available.

NE is defined as ME minus the heat increment, which is the heat produced (and thus energy used) during digestion of feed, metabolism of nutrients and excretion of waste. The energy left after these losses is the energy actually used for maintenance and for production (growth, gestation and lactation). This means that the NE system is the only one that describes the energy that is actually used by the pig. NE is, therefore, the most accurate and unbiased way to date of characterizing the energy content of feeds. However, NE is much more difficult to determine and more complex than DE or ME, which may be a reason why it is not as widely used as it should be. Currently, only France, the Netherlands and Germany have developed NE systems to describe dietary NE contents of feedstuffs, although research into NE has been conducted in several countries.

DE requirements are applied most precisely in the case of animals that are fed rationed amounts of feed, e.g. gestating sows. The DE requirement of the gestating sow varies with body weight, weight gain, environmental temperature and other factors. Assuming that the sow gains 25kg in body weight during pregnancy and gains 20kg in the form of piglets plus uterine and udder tissue, the maintenance need can be calculated as being similar to that of a growing-finishing pig and the need for gain can be calculated as 1.29Mcal DE/kg. Thus, the total requirement is 5.8–6.8Mcal DE/day for a 120–160kg animal at the start of pregnancy, equivalent to 1.8–2.0kg daily of a maize/soya diet.

The powerful anabolic effect of pregnancy can be seen in Table 3.4. Gilts in one group were bred at puberty and fed rationed amounts of feed until they farrowed. Gilts in another group were not bred at this time and were fed the same rationed amounts of feed until the bred gilts farrowed. As the results demonstrate, all the gilts received the same total amount of feed yet the bred animals were able to produce a litter of pigs while reaching the same live weight after farrowing as the non-bred animals. A litter of pigs for nothing!

	Bred	Not bred
Age when bred (puberty, days)	163	166
Weight at breeding (kg)	98	98
Weight at 85 days after breeding (kg)	116	108
Weight at 100 days after breeding (kg)	122	113
Weight after farrowing (kg)	115	118
Piglets born (n)	9.1	-
Piglets born alive (n)	8.5	-

TABLE 3.4 Growth and farrowing data for gilts bred or not bred at puberty and fed the same rationed amounts of feed/day until farrowing (from Friend et al., 1982).

These data demonstrate, in a remarkable way, the increased efficiency of the pregnant pig and explain why gestating gilts and sows need to be fed restricted amounts of feed to prevent them becoming too fat. This increased efficiency can be explained by hormonal changes in metabolism.

For the lactation stage a milk yield of 5–7.5kg/day may be assumed, also that the maintenance need is the same as for the pregnant sow. Thus, the total need is 14.5–20.5Mcal DE/day for sows of 145–185kg post-farrowing, equivalent to 4.4–6.1kg daily of a maize/soy diet.

In the case of other classes of pigs that are fed *ad libitum*, the DE requirements are met by regulating the DE content of the dietary mixture.

The main sources of energy are starch and fat. Very young weaned pigs need some simple sugars. Fibre is not well utilized and increased fibre in the diet may result in reduced digestibility of energy and protein. Although fibre cannot be digested in pigs it may be broken down to some extent by microbial action in the lower gut, providing 5–28% of the DE requirement for maintenance in the form of volatile fatty acids.

In addition to providing energy, fats supply some essential fatty acids (EFA). According to the NAS–NRC (1998), linoleic acid is the only required EFA.

The requirement tables published in the NAS–NRC (1998) report assume *ad libitum* feeding (feeding to appetite) in most cases. Where limit (restricted) feeding is used the requirements are set out as total amounts of nutrients to be consumed daily.

Owing to the need in most instances to try and obtain maximum intake of feed and in other instances to find ways of controlling feed intake, research has been conducted on the factors that control voluntary feed intake. Among the factors known to be important are diet composition. For instance, pigs will eat more of a low-protein diet than an unbalanced protein diet; fat inclusion in the diet delays stomach emptying; intake is related inversely to energy concentration in the diet; a high environmental temperature reduces appetite; and a low environmental temperature increases intake but the higher intake may be used mainly to maintain body temperature and not enhance growth. Also, it is known that castrates eat more than boars or gilts and that some breeds such as Durocs and cross-breeds eat more.

In the case of piglets and growing pigs, the digestibility and palatability of the diet, as well as the balance of dietary amino acids, have to be considered in ensuring a desirable feed intake. With finishing pigs nearing market weight it may be desirable to reduce the energy level of the diet to ensure good carcass grades.

Bulky feeds are sometimes used with gestating sows to prevent them becoming too fat. With lactating sows a high-energy feed may assist in maintaining adequate milk production when the appetite is low. It is also known that too much feed in pregnancy leads to reduced intake during lactation.

Protein

The term protein usually refers to crude protein (CP; measured as nitrogen content × 6.25) in feedstuffs composition and requirement tables.

Protein is required in the diet as a source of amino acids (AA), which can be regarded as the building blocks for the formation of muscle tissue, milk, etc. There are 22 different AA in the body of the pig, ten of which are essential (EAA; arginine, methionine, histidine, phenylalanine, isoleucine, threonine, leucine, tryptophan, lysine and valine), i.e. cannot be manufactured by the body and must be derived from the diet. Cystine and tyrosine can be regarded as semi-essential in that they can supply part of the sulphur-containing and aromatic AA requirements respectively. The other ten are non-essential (NEAA) and can be made by the body.

The most important EAA is lysine, which is usually the first-limiting AA and has to be at the correct level in the diet. The level of the first-limiting AA in the diet normally determines the use that can be made of the other EAA in the diet. If the limiting AA is present at only 50% of requirement, then the efficiency of use of the other AA will be limited to 50%. This concept explains why a deficiency of an individual AA is not accompanied by specific deficiency signs: a deficiency of any EAA results in a generalized protein deficiency. The primary sign is usually a reduction in feed intake that is accompanied by increased feed wastage, impaired growth and general unthriftiness. Excess AA are not stored in the body but are excreted in the urine as nitrogenous compounds, much of which are converted to ammonia by microorganisms in the faeces.

In most pig diets, a portion of each AA that is present is not biologically available (bioavailable) to the animal. This is because most proteins are not fully digested and the AA are not fully absorbed. The AA in some proteins such as milk products are almost fully bioavailable, whereas those in other protein feedstuffs are less bioavailable. It is therefore more accurate to express AA requirements in terms of bioavailable AA, as is done in modern conventional herds.

For optimal performance the diet must provide adequate amounts of EAA, adequate energy and adequate amounts of other essential nutrients. All of these should be of high bioavailability. Some variation in total dietary protein level can be tolerated, provided the correct levels of AA are maintained. Gilts and boars, which are leaner and eat less, need higher levels of EAA and CP in the diet than castrates which are fatter and eat more. Maximal carcass leanness requires a greater intake of EAA than does maximal growth rate.

The CP requirement values outlined by the NAS–NRC (1998) assume a maize/soya diet and have been adjusted for the average bioavailabilities of EAA in maize/soy diets. A separate table on bioavailabilities is now included in the NAS–NRC publication, allowing the dietary target values to be adjusted when diets based on other feedstuffs are formulated. Bioavailability is measured as the proportion of dietary EAA that has disappeared from the gut when digesta reach the ileum. The NAS–NRC recommends formulation on the basis of bioavailable EAA if the ileal digestibility of lysine, methionine, threonine or tryptophan is less than 70% or more than 90%.

APPLYING THE NAS–NRC ESTIMATES OF REQUIREMENT TO ORGANICALLY RAISED PIGS

The general approach to feeding organic pigs can be based on methods used in conventional production, but the regulations of organic production require several changes to be made.

The most important effect of organic production on feeding is on energy requirements. Outdoor access increases the energy needs of all classes of pigs, due to increased physical activity exercise and the need for animals to keep warm in colder climates.

Provided the necessary changes are made to the feeding regime, the performance of herds with outdoor management can be similar to that of herds housed conventionally. This was demonstrated in the results of a survey conducted in the UK (Table 3.5).

Measure	Outdoor	Indoor
Sow mortality (%)	3.1	3.9
Replacement rate (%)	45.8	47.7
Conception rate (%)	82.2	81.6
Litters per sow per year	2.19	2.25
Liveborn piglets per litter	10.9	11.4
Mortality of piglets born alive (%)	12.3	13.0
Pigs weaned per sow per year	20.9	22.4

TABLE 3.5 Performance of breeding herds with outdoor management or housed in conventional indoor facilities in the UK (from BPEX, 2008).

These results indicated that outdoor management gave slightly better health, as shown by the mortality and replacement rates. However, the number of pigs weaned per sow per year and litter size were lower with outdoor management, confirming the results of other studies. This was not due to conception rate, which was slightly better with outdoor management. Farrowing problems were also slightly reduced with outdoor management.

Performance was lower than average for both types of management, compared with a corresponding average of more than 25 pigs weaned per indoor sow per year and an average of 21.5 per outdoor sow per year for outdoor herds.

These results indicate that producers using outdoor management (including organic producers) need more advice on how to increase sow productivity.

The above report also indicated a high sow replacement rate. Sows produce larger litters than gilts so keeping the number of gilt replacements to a minimum will improve production efficiency. Most commercial producers replace 30 to 40% of their herd each year. Therefore, the nutrition and management of gilts is an important consideration in optimising reproductive efficiency.

The long lactation period (6–8 weeks) in organic production is another factor that needs to be taken into account since it can result in the sow losing condition and taking longer to re-breed.

FEEDING THE BREEDING HERD

In general the following is suggested as the minimum range of diets necessary for feeding a herd of organic pigs, assuming a farrow-to-finish operation involving all ages of stock:

- Creep feed.
- Starter diet.
- Grower diet.
- Finisher diet.
- Sow gestation diet.
- Sow lactation diet.
- Boar diet.

A key aim of organic farming is environmental sustainability. Consequently, organic producers wish to provide most or all of their required inputs, including feed. However, this is not possible on small farms, and even larger farms that may produce some of the feedstuffs required may not have the necessary mixing equipment to allow adequate diets to be prepared on-site. Farms with a land base sufficient for the growing of a variety of crops may be able to mix diets on-site or in a co-operative mill.

Creep feed is generally a complex type of feed, containing several ingredients not likely to be available on farms. Consequently, it is recommended that organic creep feed be purchased from a reliable source.

Organic standards require that pigs receive forage in their diet. This can be achieved in various ways. One way is simply to allow the pigs access to pasture. Another is to incorporate dried or dehydrated forage in the feed mix, or by allowing them *ad libitum* or restricted daily access to products such as silage or root crops together with a supplementary feed mixture to provide the necessary nutrients lacking in the forage source.

STANDARDS

There is currently no set of nutritional standards derived specifically for organic pigs. These standards have to be derived from existing standards for pigs raised conventionally, such as those derived by the US NAS–NRC that are used widely in many countries.

One criticism of the estimates of nutritional requirements published by the NAS–NRC is that they relate mainly to improved genotypes of fast-growing pigs with a high potential for lean meat development and low carcass fat content.

Most organic producers use traditional breeds of pig that have not been subjected to the selection pressure imposed on leading genotypes used in conventional production. The traditional breeds are mainly of slow-growing types, which are fatter and contain less lean meat than improved breeds. Consequently the organic pig industry should find the older NAS–NRC publications the most useful guide to nutrient requirements, although the more recent publications provide better estimates of the requirements for minerals and vitamins.

Therefore, it is logical to modify the more recent NAS–NRC values by lowering the estimated requirement for energy in pig diets to a value that is less likely to promote excessive fatness in these slower growing animals. Reducing the energy content of the

diet also minimizes the need for restricted feeding to avoid over-fatness. In reducing the energy value of the diet it is advisable to reduce the protein and limit amino acids by a similar proportion so that the same ratios of energy to protein and amino acids are maintained.

However, it is recommended that the NAS–NRC derived values for mineral and vitamin requirements are adopted without modification, to help ensure correct skeletal growth and avoid foot and leg problems.

Modification of the NAS–NRC values in this way brings the recommended standards broadly in line with organic standards suggested by other agencies and scientific groups.

Suggested standards based on this approach, and examples of feed formulas meeting the standards, are shown in the tables later in this section.

On the other hand, organic producers using modern hybrids and genotypes with a high potential for lean meat development and low carcass fat content may find the requirement values recommended by the breeding company for that particular genotype to be more useful than the values suggested here.

	Bred gilts, sows and boars	Lactating gilts and sows
DE (kcal/kg)	3000	3100
DE (MJ/kg)	12.55	13.4
ME (kcal/kg)	2900	3070
ME (MJ/kg)	12.15	13.0
CP (g/kg)	120	170
Lysine (g/kg)	5.5	8.5
Calcium (g/kg)	7.5	9.0
Phosphorus, available (g/kg)	3.5	3.5

TABLE 3.6 Recommended dietary standards for the feeding of organic breeding stock, amount per kg diet (900g/kg moisture basis).

Ingredient, kg	Example 1	Example 2	Example 3	Example 4
Barley	549	800		671
Maize			450	
Wheat		89		
Wheatfeed	260		394	
Oats				150
Soya bean meal		80	75	
Canola meal				100
Peas, field	170			
Faba beans				50
Dehydrated lucerne			50	
Ground limestone	17	12	10	10
Dicalcium phosphate	10	15	15	15
Salt (NaCl)	3	3	3	3
Vitamin Premix	5	5	5	5
Trace mineral Premix	5	5	5	5
Total, kg	1000	1000	1000	1000

TABLE 3.7 Example organic feed mixtures for bred gilts, sows and boars that meet the recommended standards.

* Nutrient composition cannot be guaranteed due to ingredient variability.

Gilts (Young Females)

During the growing-finishing phase gilts are generally fed unrestricted daily amounts of feed to market weight using finisher feed, and continue to be fed *ad libitum* or restricted to around 3kg feed per day to breeding weight if they are becoming too fat. If restricted, *ad libitum* feeding for 7–10 days prior to breeding is likely to induce flushing of ova, possibly resulting in larger litters. Ovulation rate in the gilt increases from puberty through the third oestrus. An initial step with breeding gilts is therefore to delay breeding until the third oestrus. This requires preventing the exposure of the gilts to a mature boar or relocating and mixing them with other animals in the herd.

Boars

Boars are usually fed gestation feed, in restricted amounts, to maintain correct body condition. They can be housed in pairs or small groups only if they have been together from a young age and do not show any aggression toward each other. They should not be kept in solitary accommodation or be isolated since they need to be housed where they can maintain visual and physical contact with compatible animals.

Bred Females

After mating, the feed allowance has to be reduced over 7 days to about 2.5–3.0kg of a gestation diet per day otherwise the animal puts on too much fat during the gestation period.

Over-fatness is undesirable for several reasons. The animal is clumsy and may overlay and crush some piglets during and after farrowing. Also, over-feeding during gestation is known to reduce feed intake during the lactation period when the sow needs to eat as much as possible to milk well and to prevent loss of body condition. Some producers increase feed intake or add fat to the diet 10–14 days prior to farrowing in an attempt to increase milk production.

Practical studies have shown that a diet containing 110–130g/kg protein and based either on barley or maize and soya bean meal or canola can be used successfully with breeding sows.

In general, energy and/or feed consumption by pregnant sows affects mainly maternal weight gain, to a lesser extent piglet birth weights and has little or no effect on number of pigs born. Using research data it can be calculated that on average it takes 114kg of feed for a pregnant gilt or sow to increase average birth weight of the piglets by 20 or 50g. This increase is not economical, especially if an average birth weight of over 1.3kg is being achieved and any increase does not improve piglet survival rate.

A good approach with pregnant sows is to allow an increase in maternal weight gain during pregnancy of around 25kg for piglets and foetal tissues and 10kg net weight gain (for the first 5 litters – zero weight gain thereafter). For sows of different weights, this translates to about 2.5–3.5kg per day of a barley–soya bean meal diet. This target feed allowance is based on favourable climatic conditions. A drop in environmental temperature from the ideal of 18–19°C by 10 degrees would require the ration to be increased by 300–600g/day.

Outdoor activity increases the energy requirement but this increase is difficult to estimate in practical situations. A reasonable approach is therefore to use the target value for weight increase as a guide and adjust the feed ration up or down to achieve it.

Producers now use condition scoring to ensure that sows and gilts are maintained in the correct body condition and are not allowed to become too fat or too thin. This helps to ensure that they are being fed correctly and will reproduce successfully. The scoring technique involves the visual assessment of bodily fatness and by light pressure on the skin over hip bones.

A common scoring guide is set out in the table below.

Score	Appearance from rear	Description
1	Emaciated	Very thin, hips and backbone very prominent, no fat cover.
2	Thin	Hip bones and backbone easily felt with no pressure.
3	Ideal	Firm pressure required to feel the hip bones and backbone.
4	Fat	Firm pressure required to feel the hip bones and backbone.
5	Obese	Visually very fat, impossible to feel bones with pressure from a finger.

TABLE 3.8 Condition scoring guide for sows.

The aim should be to have the sows scoring in range 3.

A body condition score of 3 on a scale from 1–5 (similar to the condition scoring system for sows) is considered desirable for breeding boars.

It is advised that sows should not reach the farrowing stage with a condition score of less than 3 and that the score should not be allowed to fall to less than 2.5 during lactation.

Generally sows are condition scored at weaning, mating, mid-gestation and at farrowing. Owing to the long weaning period required in organic production it is also recommended that condition scoring be conducted midway through the weaning period and when the litter is weaned. Body condition lost during lactation needs to be regained during gestation.

Grazing

Organic production requires that the animals be provided with outdoor access, a system that is perceived by consumers as being a type of farming that is more sustainable, traditional and family based.

Allowing dry sows to have access to grazing allows them to express their natural behaviour and it provides the animals with part of their nutritional needs. Legumes such as clover and lucerne are preferred for grazing because of their higher nutrient content. A rotational grazing system should be planned, so that the forage is available at all times during the year when the weather permits. Applying nose rings to the sows will prevent rooting and damage to the pasture, while still allowing grazing. However, the use of nose rings would have to be approved by the local certifying agency.

Sows are usually managed in groups during this period and need to be fed in ways to avoid bullying and to ensure the correct intake of feed individually.

It is advisable to continue to allow the dry sows access to gestation feed so that they maintain proper condition and produce healthy litters. The feed ration can be reduced when the sows are on good pasture, provided there is no reduction in condition score. The intake of minerals and vitamins needs to be maintained, however. For instance, if the feed allowance is reduced to half, then the amounts of calcium, phosphorus, salt, trace minerals and vitamins added to the diet should be doubled.

During the spring growth flush, the pasture should be used efficiently. The optimum pasture height is 10–12cm. At this time the stocking rate can be increased by increasing the density of animals in the pasture area or by dividing it up. Mowing pastures with excess growth will maintain the quality of the forage.

Rotation of pastures or dry lots is recommended to prevent the build-up of parasites, their eggs being shed in the manure. These eggs can remain viable for a considerable period and without rotation the animals will be constantly infected with internal parasites.

Lactation

During lactation the approach should be to increase the intake of feed to *ad libitum* using a diet that is higher in energy and protein than the gestation diet.

The energy and protein requirements of the lactating sow will depend on the weight of the sow, her milk yield and its composition. In general, a 170g/kg protein diet is satisfactory for lactating sows fed either barley or maize-based diets. Recommended standards for lactation diets are shown in Table 3.6.

For several reasons (breed, strain, environmental temperature, level of feed fed in gestation and palatability of diet) a sow may eat less than the target intake of 5.5–6kg per day. Research findings indicate that energy intakes lower than 50MJ of ME/day are detrimental to sows returning to oestrus, with fat loss being a major contributing factor. It is also known that inadequate protein intake during lactation will delay return to oestrus.

Increasing the level of feed post-weaning may shorten the interval to service,

Ingredient, kg	Example 1	Example 2	Example 3	Example 4
Barley	49	26		0
Maize		300		200
Canola meal			100	100
Wheat	350	337	585	320
Wheatfeed	300	75	100	200
Soya bean meal	100	230		100
Fishmeal (white)			25	
Peas	170		168	25
Faba beans				50
Ground limestone	17	15	15	14
Dicalcium phosphate	10	13	4	12
Salt (NaCl)	3	3	2	3
Vitamin Premix	5	5	5	5
Trace mineral Premix	5	5	5	5
Total, kg	1000	1000	1000	1000

TABLE 3.9 Example organic feed mixtures for lactating gilts and sows that meet the recommended standards.

* Nutrient composition cannot be guaranteed due to ingredient variability.

especially in sows that have lost condition during the lactation period.

Weaning should last as long as possible and at least for 40 days. This has a beneficial influence on the piglets but places a greater strain on the sow in terms of an extended period of milk production and possible loss of body condition. It can also result in impaired fertility and lower overall productivity. Lactating sows should be allowed access to a self-feeder along with grazing to maintain proper body condition.

Outdoor straw-bedded ark systems on pasture work well during this stage. These provide the sow with a great opportunity to express maternal behaviour such as nest-building and provision of environmental stimuli to the progeny. Providing outdoor conditions also allows the piglets to mix with other litters prior to weaning, avoiding later problems of aggression and bullying.

Another major European outdoor system is the traditional Mediterranean silvopastoral system. This system uses indigenous breeds that are pastured extensively in natural forests for the production of high-value, dry-cured hams. Typically, all phases of production take place outdoors, sometimes in extreme conditions in mountain zones. These animals reach their slaughter weight at an advanced age, conferring preferred attributes to the meat. In addition, finishing takes place during autumn in forests of oak or chestnut and the animals convert large quantities of acorns or chestnuts into fat deposits. There is thus no standard outdoor system for pig production in Europe. A similar situation prevails in other regions.

An outdoor system based on the 'Roadnight' system employing movable arks (nicknamed 'pigloos') that gave good results in the rigorous climate of north-east

	Roadnight	Conventional
Number of farms	9	9
Average amount of feed		
Purchased Meal/Cube (tonnes per sow per year)	1.15	1.30
Skimmed milk (litres per sow per year)	0	414.1
Labour		
Average (hours per sow per year)	18.2	35.8
Average number of veterinary calls (per 10 sows per year)	0.9	8.2
Weaners produced		
Average number of litters (per sow per year)	2.0	1.9
Average number of piglets weaned (per sow per year)	17.4	16.7
Average piglet weight at 8 weeks (kg)	20.1	17.5

TABLE 3.10 Efficiency of weaner production by the Roadnight and a conventional pig-rearing system in Aberdeenshire, Scotland (from Blair and Reid, 1965).

Scotland was described by Blair and Reid (1965). These researchers found that the system was, on a per sow basis, at least as efficient as a conventional system of rearing. Some figures indicating the physical and economic efficiency of the two systems are given in Table 3.10.

In this system groups of sows were batch-farrowed twice yearly during March and September in small individual wooden arks on pasture, spaced at 6–8 per hectare. The piglets remained with their dams up to the age of 8 weeks when they were disposed of as weaners or were transferred to finishing yards. At this time the sows were dried off, re-bred and housed outside in larger communal arks until the next farrowing. Generally only cross-bred sows were used for the system, usually between Wessex Saddleback and Large White or Landrace. This cross was selected to combine the traditional hardiness and mothering ability of the Wessex Saddleback with the better carcass characteristics of the latter breeds. The sow feed was generally a purchased feed in either cube or nut form, which was considered more suitable than a meal for feeding to stock on pasture. Creep feed was generally in pellet form and given to the young pigs from the age of 2–3 weeks in covered feeders. Drinking water was supplied to each ark from a central point by the use of alkathene tubing, which had the advantage that the water system rarely froze.

Sows on the Roadnight system produced on average two litters of heavier piglets per year. No information was available on the carcass quality of the weaners produced on this system since they were marketed to other farmers for finishing. The slightly lower number of weaners produced on the conventional system may be attributed to a common difficulty in getting sows to conceive in the late autumn in the north-east of Scotland, especially when housed conventionally. Total feed costs were slightly lower on the Roadnight system. Although the amount of feed used was considerably less than on the conventional system (due probably to the provision of pasture), the increased cost of purchased feed reduced the potential savings on the Roadnight system. Savings in the costs of feed for this type of operation could be obtained by group trading and access to a mill for cubing, and by the use of home-grown grain in the diets. Housing costs were lower with the outdoor system. An important finding was that the amount of veterinary attention needed for these herds was much less with the outdoor system. The Roadnight system did demonstrate one disadvantage of outdoor systems in colder climates, namely the concentration of weaners at two periods in the year.

FEEDING YOUNG PIGS

The pig is born with about 47mg iron in body stores, sufficient for about 1 week. Sow's milk is very low in iron (about 1mg/kg), therefore, unless the pig is given extra iron it will become anaemic as early as 5–10 days of age. Thus, iron supplementation is very important for the baby pig, to prevent symptoms of poor growth, listlessness, rough hair coat, etc. A characteristic sign of anaemia is laboured breathing after minimal activity ('thumps').

Organic producers usually provide sods of earth in the farrowing pen for the baby pigs to root and obtain a supply of iron. It is important that the sods be free of disease organisms and parasites. Iron supplementation of creep feed and starter diets is necessary for piglets raised outdoors in yards. Diets fed to older animals usually contain adequate levels of iron.

Milk production of the sow depends on the individual, lactation number, litter size and stage of lactation. After about 3 weeks, milk production starts to decline and there is a nutrient deficit period unless creep feed is provided to allow the young pig to adapt to solid feed and continue to grow well. The stomach capacity of the young pig is low, therefore a diet of high nutrient density should be used. Creep feed has to be formulated carefully from highly digestible

ingredients since at an early age the piglet gut has not yet adapted to non-milk diets. Indigestible feed is likely to result in digestive upsets and scouring. Consequently, it is recommended that producers purchase creep feed rather than attempt to mix it on the farm.

This diet should contain 200–240g protein/kg (11–14g lysine/kg) and include milk products among its ingredients. Some producers sprinkle dry powdered milk on the creep feed to improve its smell and palatability. This diet may be offered as small pellets or crumbles to stimulate consumption.

Social interaction between the piglets while eating is important in developing feeding behaviour. A well-designed feeder without solid partitions encourages good social interaction and maximum feed intake, yet preventing the small pigs from lying and dunging in the feeders.

After about 5–6 weeks the piglets should weigh about 12–15kg and can be gradually switched over to a starter diet. This diet can be mixed on the farm and is more economical, being lower in nutrient content and in digestibility than creep feed.

The starter diet introduced while the piglets are still suckling should be continued after weaning until the piglets have reached an average weight of about 25kg, then mixed with grower diet for a few days before switching completely to a grower diet.

Feeding Weaned Pigs

Weaning is a stressful event for the young pig, commonly resulting in a 'post-weaning lag' when it fails to grow for some time and may even lose weight. However, by 40 days of age it should be possible to wean the pigs successfully, since the stress of weaning diminishes as they become older. Other strategies that help include moving

	Creep feed	Starter feed
	Piglet weight 5–15kg	Piglet weight 15–25kg
DE (kcal/kg)	3400	3300
DE (MJ/kg)	14.25	13.8
ME (kcal/kg)	3250	3170
ME (MJ/kg)	13.65	13.25
CP (g/kg)	200–240	190
Lysine (g/kg)	11–14	9.5
Calcium (g/kg)	11	7
Phosphorus, available (g/kg)	3.5	3.2

TABLE 3.11 Recommended dietary standards for the feeding of organic piglets, amount per kg diet (900g/kg moisture basis).

Ingredient, kg	Example 1	Example 2	Example 3	Example 4
Wheat	540	388	488	410
Wheatfeed	100		189	200
Maize		255		100
Soya bean meal	279	325		
Full-fat soya beans			100	150
Faba beans				50
Peas, field	50		135	
Fishmeal, white			75	75
Ground limestone	18	17	11	10
Dicalcium phosphate	9	11		3
Salt (NaCl)	3	3	1	1
Vitamin supplement	0.5	0.5	0.5	0.5
Trace mineral supplement	0.5	0.5	0.5	0.5
Total kg	1000	1000	1000	1000

TABLE 3.12 Example Starter feed mixtures that meet the recommended standards.

the sow out of the farrowing area (rather than the reverse) and leaving the young pigs there for a few days before moving them to the rearing accommodation. Having the pigs eating sufficient starter feed daily will ensure they are accustomed to solid feed at weaning. Also, the feed should be attractive and of high nutrient quality to ensure high digestibility. Once the pigs are eating successfully they can be introduced gradually to a grower diet, which does not have to be pelleted if meal-type diets will be fed during the growing and finishing stages. The feeder should be of a design that allows easy access by the pig and producer, stays relatively clean and can be placed securely inside the creep area. Any feeders contaminated with urine or faeces should be replaced at the next feeding with a clean feeder. The natural inclination of a pig is to root or seek feed on a flat surface and most feeders are designed to accommodate this behaviour.

Careful attention needs to be paid to hygiene and avoidance of disease at this time. Digestive upsets immediately post-weaning increase susceptibility to disease. The pen needs to be warm and draught free.

The most suitable post-weaning diet is one containing feed ingredients that are palatable, highly digestible and low in anti-nutritive factors. The diet should have a moderate protein content of high quality and a high content of digestible energy. Digestibility of fat, protein and carbohydrate in the starter diet are important to avoid digestive upsets while the gut adapts to a dry diet. Some producers feed the diet as gruel for the first few days to ease this transition. Very high-protein diets (>200g/kg CP) may increase digestive upsets.

The correct choice of sources of fibre and complex carbohydrates helps to ensure the functional development of the digestive system and the gut microflora. These include oligosaccharides and wheat fibre ingredients. Sow's milk is very high in fat (more than 300g/kg dry matter), yet commercial starter diets with similar fat levels are likely to produce digestive upsets. This is probably related to the fatty acid composition of the fats used.

Improved feed intake immediately post-weaning can be achieved by the use of liquid or gruel feeding as opposed to dry feeding. Also, an adequate intake of clean water is important in helping to ensure an adequate feed intake.

Feeding Market Pigs

The requirements for energy and protein change as the animals grow, the requirement for protein in relation to energy decreasing over time. As a result, it is advisable to use several diets over the period from weaning to market at around 100kg live weight. It is common to use a three-stage feeding programme: i.e. starter diet from weaning to about 45kg, grower from then until the pigs reach about 50–60kg, followed by finisher from then until market. Usually these diets are fed *ad libitum* (unrestricted amounts) for maximum efficiency. Pigs that are likely to be too fat at market because of their genotype (especially castrates) should be placed on a restricted scale of feeding during the finisher stage. Penning by sex may have to be used to achieve this objective. Gilts usually grow more slowly than castrates because they eat less, particularly above 50kg body weight. They generally deposit the same amount of lean tissue per day as castrates but deposit less fat because of their lower energy intake. Gilts are therefore somewhat older and leaner by the time they reach market weight. Consequently, because of the lower feed intake but a similar development of lean tissue, gilts require a higher concentration of dietary amino acids than castrates. It is desirable, therefore, to feed gilts and castrates in separate pens.

	Grower	Finisher
	Weight of pig 45–70kg	Weight of pig 70kg to market
DE (kcal/kg)	3100	3050
DE (MJ/kg)	13.0	12.75
ME (kcal/kg)	2975	2925
ME (MJ/kg)	12.45	12.25
CP (g/kg)	160	140
Lysine (g/kg)	7.5	7.0
Calcium (g/kg)	6.0	5.0
Phosphorus, available (g/kg)	2.3	1.9

TABLE 3.13 Recommended dietary standards for the feeding of organic growing-finishing pigs, amount per kg diet (900g/kg moisture basis).

Ingredient, kg	Example 1	Example 2	Example 3	Example 4
Barley	545	735	312	
Wheat	45			
Maize			120	
Sorghum				400
Canola meal			100	
Soya bean meal	135	130		100
Fishmeal white				25
Peas	55			75
Faba beans			200	
Wheatfeed	200	115	250	386
Ground limestone	11	12	12	8
Dicalcium phosphate	5	4	2	3
Salt (NaCl)	3	3	3	2
Vitamin supplement	5	5	5	5
Trace mineral supplement	5	5	5	5

TABLE 3.14 Example organic feed mixtures for growing pigs that meet the recommended standards.

Ingredient, kg	Example 1	Example 2	Example 3	Example 4
Barley	450	735	397	
Wheat				
Maize			120	76
Sorghum				379
Canola meal			120	
Soya bean meal		130		100
Fishmeal white				50
Peas	250			
Faba beans			120	
Wheatfeed	280	115	225	387
Ground limestone	11	12	12	5
Dicalcium phosphate	5	4	2	1
Salt (NaCl)	3	3	3	1
Vitamin supplement	5	5	5	5
Trace mineral supplement	5	5	5	5

TABLE 3.15 Example organic feed mixtures for finishing pigs that meet the recommended standards.

Split-sex feeding allows for the more precise formulation of diets to meet the nutrient needs of castrates and gilts. Also, gilts may exhibit oestrus in the finisher pen before they reach slaughter weight, which can affect the feed intake and growth of the other animals. Separation of pigs by sex to different pens at weaning or at the start of the grower period allows the producer to market leaner gilts and to feed castrates more economically. It also results in pigs of a more uniform weight within each pen.

Ad libitum feeding from self-feeding hoppers has certain advantages. It reduces labour costs, makes feeding management somewhat simpler and maximizes growth rate. However, there are disadvantages.

If pigs tend to deposit fat early, carcass grades will be reduced. Maximum intake of feed will be reduced if feed is allowed to become stale or is fouled, or if trough space is inadequate and causes competition for the available feed. Therefore good management is required to eliminate these problems. When self-feeders are used, care must be taken to ensure they are adjusted properly to prevent waste.

Young growing pigs use feed economically. They should be encouraged to eat as much as possible. The breed of pigs being raised, previous market grades and production costs will determine whether restricted feeding may be more economical once pigs reach about 60–70kg live weight.

Energy requirements of outdoor pigs are generally higher because of increased exercise and exposure to outside temperatures, but protein requirements are relatively unaffected.

Most organic producers, at least in the temperate regions, do not make changes in the composition of the diet to account for winter temperatures. During cold weather

animals fed *ad libitum* simply eat more to maintain body processes and animals fed restricted amounts of feed have to receive a higher feed allowance. The ideal environmental temperature for a growing pig is around 18–20°C and for a finishing pig is slightly lower, about 16°C. If the temperature falls below the ideal it is advisable to increase the amount of bedding in the sleeping area as a way of helping to keep the animals warm. Producers accept higher feed intakes and lower efficiencies of feed utilization during cold weather. It is logical also to increase the allowance of forage at this time. The increased intake of fibre results in an increased heat of fermentation in the gut, which helps to keep the animals warm. However, as overall efficiency of organic production becomes more important, producers may have to alter the composition of dietary mixtures for animals with access to outdoors during periods of cold weather. It would be logical to reduce the concentration of protein and amino acids to take into account the increased intake of feed, and maintain the target intake of nutrients on a daily basis. For instance, if intake increased by 10% then the concentration of protein and amino acids could be reduced by about 10%. Another approach could be to raise the energy level of the diet in relation to protein and other nutrients, perhaps by the use of fat, in such a way as to result in the same intake of amino acids but an increased intake of energy. Such changes should be made following advice from a nutritionist. Few organic producers are in a position to make these changes, although those using purchased feed could do so in collaboration with a feed manufacturer. As noted above, during cold weather it is advisable that the animals be provided with ample, dry bedding in the sleeping area.

BREEDS FOR ORGANIC PRODUCTION

The type of pig selected for organic production needs to do well under organic conditions and produce meat of high quality at an acceptable price. Where acceptable under the local organic regulations, a pig of improved type (hybrid) should be first choice. Alternatively, producers may wish to use heritage breeds.

Using heritage breeds allows producers to market the meat as 'heritage', 'original' or 'traditional'. These breeds are likely to be better adapted to the region and to produce meat of better eating quality, but are likely to be unimproved in terms of growth performance and carcass quality. Therefore, their meat may be produced at high cost.

An example of a heritage breed is the Portuguese Bísaro pig, with meat of excellent quality and suitable for processing. However, it has a slow growth rate, bad conformation and excessive adult weight and body size.

It is desirable for pigs to be cross-bred in order to obtain the full advantage of heterosis (hybrid vigour). In conventional production the sow is commonly a cross-bred (F1 generation) obtained from a crossing of selected animals of two pure breeds (e.g. Yorkshire (also known as Large White) × Landrace). These cross-bred sows are then bred to selected boars of a third breed such as Duroc or Hampshire to provide additional heterosis and desirable carcass characteristics in the progeny (F2 generation) to be marketed as meat animals. The F2 animals do not enter the breeding herd. Surveys show that cross-bred sows farrow larger pigs with an increased rate of survival. The pigs are heavier at weaning and make more rapid gains from weaning to market.

The main objectives of a cross-breeding policy are to maximize three traits: sow productivity (number of pigs weaned or sold per sow per year), efficiency of gain (growth rates and feed conversion ratios) and suitability of the carcass for the selected market (weight, length, fat depth/quality, skin and meat and fat colour). Only large organic units would be capable of producing this type of stock on a regular basis. A compromise would be to purchase boars and cross-bred gilts as required or to use artificial insemination, within the limits imposed by the local organic regulations.

Producers using a closed herd policy will find a benefit from a reduced disease risk, which is higher in herds that allow replacement stock to be bought in. However, a disadvantage of a closed herd policy is the real threat of inbreeding and a resultant loss of vigour in the herd.

The traits important for a good outdoor sow include prolificacy, good conformation, strong vigour, good maternal behaviour including ease of handling, and adequate fat reserves in northern regions to provide against cold conditions. The sows used in Europe for outdoor production are commonly Saddleback, Landrace, Large White and Duroc crosses. Traditionally sows in the UK were Saddlebacks because of their grazing ability and hardiness. During the Second World War about half of all sows were of this breed, which were crossed with a white boar to produce a dual-purpose pig for combined pork and bacon production.

Durocs have the same characteristics of hardiness and good mothering ability as Saddlebacks but have significantly better growth and carcass characteristics. Boars need to be able to withstand the outdoor conditions as well, and to have the right temperament for group management. For the conventional market, boars are normally selected for lower backfat, to counter the backfat traits in sows. Conventional outdoor systems may use Duroc, Large White or even Sire Line boars to provide the traits for lean meat.

Despite improved breeding and selection, the progeny from outdoor pig production systems in Europe tend to have increased backfat compared with equivalent pigs from indoor units, probably in response to the outdoor environment. This makes the choice of sire line very important.

Coloured breeds such as Durocs and Hampshires have greater resistance to sunburn than entirely white breeds, such as the Landrace and Large White. However, shade can be provided to mitigate the effect of sunburn.

Saddleback, Duroc and the Camborough 12 are recommended for organic production by some organic certifying agencies.

Organic producers are encouraged to use traditional breeds that may be more suited to local conditions than improved genotypes. A large number of these breeds exist in several countries, though often in relatively low numbers. One disadvantage of using a pure breed on small farms is an inadequate size of the breeding herd, leading to the problem of inbreeding that results in loss of productivity in the stock.

Traditional breeds are currently well represented in smaller organic farms in the UK (ADAS, 2001), e.g. British Saddlebacks. Other examples are Gloucester Old Spots and Tamworths, both of which have the reputation of being hardy and excellent mothers. Large Blacks, Berkshires and Wessex Saddlebacks are being used in Australia. Suggestions from some organic producers are that a Saddleback × Duroc cross-bred sow provides hybrid vigour and overcomes the Saddleback's disadvantages of lower litter sizes, increased carcass fatness and reduced feed conversion efficiency. Saltalamacchia et al. (2004) suggested that Siena Belted, Casertana, Romagnola,

FIGURE 3.2 Tamworth pigs, suitable for organic production. (Photo: iStock)

Calabrian, Black Madonie, Duroc, Large White × Duroc and Large White × Siena Belted stock are suitable for organic production in Italy. Uremovic et al. (2003) found that Black Slavonian and F1 crosses with Duroc were suitable for European continental conditions and that German Landrace were suitable for Mediterranean conditions. Climate is therefore an important factor influencing breed selection for outdoor-based organic production.

The benefits of cross-breeding in sows were demonstrated by Gaugler et al. (1984). Data on litter size and weight at 42 days and piglet survival to 42 days were collected from 366 litters of Duroc, Yorkshire, Spotted and Landrace sows, either pure-bred or cross-bred. Results were as follows for litter size, litter weight (kg) and survival to 42 days (%) respectively: Duroc 6.29, 70.30 and 67.80; Spotted 6.85, 74.26 and 74.25; Yorkshire 7.25, 73.53 and 64.21; and Landrace 7.88, 89.79 and 76.97. When cross-bred the average advantage over the pure-breds at 42 days was 0.79 more pigs, 11.72kg extra litter weight and a 5.56% greater survival rate. The best overall productivity was from Duroc × Landrace sows, with litters averaging 97.99kg at 42 days and a survival rate of 83.71% from birth.

Kelly et al. (2005a) investigated the effect of breed on sow performance in organic systems in the UK. The study compared maternal performance of three breed types: 'traditional' (Saddleback; S), 'cross-bred traditional' (Saddleback × Duroc; SD) and 'modern' (Camborough 12 hybrids; C12). Twenty gilts of each genotype were used on each of two commercial organic farms. They received standard feeding (no details) and management according to the farm's normal commercial practice. All sows were mated using a same group of Duroc boars, with individual sires balanced across genotypes. Reproductive performance and health

were recorded over four pregnancies. No genotype showed a greater incidence of health or welfare problems when managed to organic standards, with a similar proportion of sows successfully completing four pregnancies (88–90%). Mean total litter size at birth was significantly greater for the hybrids (C12=11.5, S=10.7, SD=10.4), as was the number of stillborn piglets per litter (C12=0.8, S=0.3, SD=0.6). Total piglet losses during lactation were greatest for the more prolific, modern genotype (C12=1.9, S=1.2, SD=1.5). The net effect on the number of piglets weaned at 60 days of age favoured the traditional Saddleback (C12=8.4, S=9.1, SD=8.0), but this was offset by a slightly longer farrowing interval (C12=194, S=200, SD=185), resulting in similar annual piglet output in the three genotypes. The authors concluded that the results confirmed the greater prolificacy of the modern hybrid and the better maternal characteristics of the traditional Saddleback. They also concluded that all genotypes could be successful in organic production systems, and that choice of sow breed should depend on the ability of the farm to manage prolific sows.

In a related study the same researchers (Kelly et al., 2005b) investigated the effect of breed and husbandry system on the growth and carcass quality of organic growing pigs. The study compared performance and carcass quality of progeny from three maternal breed types that were kept either on pasture or in housing with an outdoor run and offered grain-based feed (no details) *ad libitum*, either alone or with fodder beet or grass/clover silage as additional forage. The sow breeds were 'traditional' (Saddleback; S), 'cross-bred traditional' (Saddleback × Duroc; SD) or 'modern' (Camborough 12; C12), and were all bred to Duroc boars. The pigs were weaned at 30kg and grown to a slaughter weight of 90kg. Live weight gain, feed intake and the proportion of intake comprised by the forages did not differ among genotypes. Carcass backfat thickness (P2) was lowest for C12 progeny, and highest for the progeny of the traditional pure breed (C12=11.4, S=14.3, SD=13.4). Intake of the forage supplements was very low when feed was available *ad libitum*. Over the whole period it averaged only 0.12kg/day for the silage and 0.16kg/day for the fodder beet, representing less than 2% of total DM intake. Growth rate did not differ between housing systems, but feed intake was increased in outdoor animals (OUT=2.47, IN=2.22kg meal equivalent per day), resulting in a significant deterioration in feed conversion efficiency. Animals allowed access to grazing consumed a smaller proportion of their daily intake from the additional forages, and had a higher killing-out percentage.

Hence the breeding programme should aim at producing breeders and progeny that are suited to outdoor conditions, requiring the stock to be hardy. The Duroc breed remains one of the top choices in North America for a terminal sire line to use on sows that are crosses between the white breeds. As the sire breed is different from the maternal breeds, hybrid vigour is at its maximum for the market hogs. This has a positive effect not only on growth rate but also on many other factors such as hardiness and resistance to disease. Hair colour should not be an issue for this type of mating. When crossed to F1 sows from Yorkshire × Landrace breeds, Durocs almost always yield white market hogs since colour is a recessive trait.

Organic standards, in Europe at least, require the implementation of a closed herd policy. Consequently, only 20% of stock on an annual basis may be bought in and only if the stock comes from non-intensive units. Since it is usual for sows to produce only about eight litters before being culled, a replacement level of 30–40% per year

is often necessary to maintain the correct balance of young to old sows.

As discussed above, it seems clear that only large organic units would be capable of producing superior cross-bred stock in sufficient numbers on a regular basis. A compromise would be to purchase boars and cross-bred gilts as required under the prevailing rules. Purchase of replacement breeding stock is therefore likely to remain a routine practice, particularly in smaller units. Another approach would be to use artificial insemination, where permitted under the local certifying regulations. Some agencies recommend that artificial insemination be avoided but its use is justified if the alternative is a marked degree of inbreeding, which is quite common in small (particularly closed) herds.

HEALTH

Good husbandry practices, including appropriate choice of breeds, housing conditions, space allowance, sanitation practices and prompt treatment of diseases, will result in a high level of animal health. Feeding the animals well is a first step since this helps the pigs to develop an active immunity to diseases.

Other steps are necessary. These include restricting visitors (human and animal) to the farm and buying in a minimum of replacement stock or perhaps purchasing no new stock and using artificial insemination (where permitted) to introduce new genes into the herd.

Another part of the programme should be to ensure that the stock is not exposed to diseases spread by rodents.

Some producers follow an all in/all out cycle, which can be very beneficial to the health status and growth performance of swine. This procedure works best when pigs are born over a short time space, and is important for successful pasture farrowing because it avoids cross-suckling of older and newborn pigs.

These steps restrict, or eliminate, the introduction of disease organisms from the environment.

Although organic pigs with outdoor access or housing are theoretically exposed to a greater disease risk, studies conducted in several countries have shown no consistent difference between conventional and organic pig production systems in the incidence of bacterial food-borne pathogens, including *Salmonella enterica*, *Campylobacter jejuni*, *Campylobacter coli*, *Listeria monocytogenes*, and *Yersinia enterocolitica* (Rostagno, 2011).

Vaccination against specific diseases that are endemic in the area may be advised. In general, vaccinations are only allowed when the targeted diseases (such as porcine parvo virus and erysipelas) are communicable and cannot be controlled by other means. Many non-organic pig herds routinely vaccinate for erysipelas and parvo virus, but this may not always be necessary. If a farm chooses not to vaccinate for erysipelas the herd should be monitored by routine blood sampling and checked for erysipelas arthritis.

The Canadian regulations are typical of organic standards in requiring that when pigs become sick or injured in spite of preventive measures they must be treated and isolated. Should the preventive practices and veterinary biologics be inadequate in preventing or treating the disease, the well-being of the animal must be protected. All appropriate medications, including the use of drugs that are not acceptable to organic production, must be used. Pigs treated with prohibited substances, such as synthetic antibiotics, at any stage of the production must be removed from organic herds or permanently identified, and cannot be

marketed as organic pork. Treating pregnant sows with medications or veterinary biologics within the first two-thirds of pregnancy may be allowed, however, the meat from these treated animals cannot be marketed as organic nor can the sows be sold as organic breeding stock.

Incidence of soil-associated problems – erysipelas, tetanus, Clostridial enteritis (*C. perfringens* type C), necrotic hepatitis (another Clostridium infection), mastitis and ringworm may all be increased with outdoor production but none of these are particularly common (unless owners fail to vaccinate against erysipelas). Lungworm (Metastrongylus) is one infection that is generally only seen when pigs have access to outdoors.

Organic producers using pasture and other outdoor systems need to rotate the use of pastures and yards to prevent contamination with external and internal parasites. Many herds have internal parasites. As the eggs of these internal parasites (worms) persist in soil for years, rotating pastures and yards may not be totally effective. Checks on the possibility of parasitism in the herd can be made by post-mortem examination of dead pigs, and by faecal sampling, slaughterhouse checks, and blood tests. Research findings confirm the likelihood of a higher incidence of infection from parasites in organic pigs. One such study found that organically reared pigs had significantly more pathological damage in the liver and lungs, caused mainly by infections with *Ascaris suum*. Also, the lungs often showed atelectasis as a result of chronic pneumonia.

According to a recent report, the most common health problems that occur in Greek organic pig farms are respiratory problems, gastrointestinal problems, claw and skin problems, parasitic infections and increased piglet mortality especially during the winter months. Measures to

FIGURE 3.3 Pigs on range. (Photo: iStock)

control these problems include improved animal welfare, the use of alternatives to antibiotics (prebiotics, probiotics and phytogenics), use of antiparasitics, appropriate vaccinations (e.g. against *E. coli*, *M. hyo*, PRRSV) and an appropriate disinfection programme (e.g. rodent management with rodenticides).

A high incidence and severity of joint condemnations at slaughter has been reported in organic free-range pigs in Sweden, compared with pigs raised in conventional indoor housing. This was confirmed by research, suggesting that modification of the housing system and breeding for joints that are more adapted to free-range movement may be needed in some types of free-range pig production. It was also found that the condemnation rate at slaughter underestimated the actual frequency of joint lesions and gave a poor assessment of joint health.

REFERENCES

ADAS (2001) Agricultural Development and Advisory Service UK Ltd, Wolverhampton, UK; available at http://orgprints.org/8122 (accessed September, 2015).

Blair, R. (2007). Nutrition and Feeding of Organic Pigs. CAB International, Wallingford, Oxford, UK. 322 pp.

Blair, R. (2012). Organic Production and Food Quality: A Down to Earth Analysis. Wiley-Blackwell, Oxford, UK. 283 pp.

Blair, R. and Reid, I.M. (1965). Outdoor Rearing by the 'Roadnight' System in NE Scotland. Agriculture 72: 530–533.

BPEX (2008). The performance of conventional indoor and outdoor breeding herds in the UK. AHDB Pork, Kenilworth, Warwick CV8 2TL, UK.

Friend, D.W., Elliot, J.I., Fortin, A., Larmond, E., Wolynetz, M.S. and Butler, G. (1982). Reproductive performance and meat

Ingredient	Maximum inclusion rate (g/kg diet)			
	Starter	Grower-finisher	Gestation	Lactation
Barley	100	800	900	500
Beet pulp	0	50	500	100
Brewer's dried grains	0	100	100	100
Cassava	0	400	400	400
Canola meal	0	150	150	150
Canola, full-fat	50	100	0	0
Cottonseed meal	0	100	150	0
Faba beans	0	200	200	100
Fishmeal, menhaden	50	50	50	50
Fishmeal, white	50	25	25	50
Flaxseed	0	50	50	50
Grass meal	0	20	50	10
Herring meal	50	25	25	50
Hominy feed	0	700	700	700
Lucerne meal, dehydrated	0	0	50	0
Lucerne meal, sun-cured	0	0	40	0
Kelp (seaweed) meal	0	25	25	0
Lupin seed	0	150	150	100
Maize	600	800	900	800
Maize gluten feed	50	100	150	50
Molasses	20	40	20	40
Oats	50	150	500	100
Oats groats	100	0	0	0
Peanut (groundnut) meal	50	75	50	150
Peas	75	300	300	200
Potato protein concentrate	50	50	50	50
Rice bran	50	150	250	150
Rye	0	25	25	10
Safflower meal	25	50	50	25
Sesame meal	25	50	50	25
Skimmed milk, spray-dried	50	0	0	0
Sorghum (milo)	400	800	900	800
Soya bean meal	150	250	150	200
Soya beans, full-fat, heat-treated	0	200	100	100
Sunflower meal	0	200	100	100
Triticale	100	500	400	400
Wheat	500	500	300	400
Wheat bran	0	100	300	100
Wheat middlings	50	400	600	200
Wheat shorts	10	400	600	200
Whey, dried	40	150	50	50
Yeast, brewer's dried	50	100	100	100

APPENDIX TABLE 3.16 Suggested maximum inclusion rates of feedstuffs in organic pig diets.

production of gilts bred at puberty and restricted in feed intake. Canadian Journal of Animal Science 62: 877–885.

Gaugler, H.R., Buchanan, D.S., Hintz, R.L. and Johnson, R.K. (1984). Sow productivity for four breeds of swine: purebred and crossbred litters. Journal of Animal Science 59: 941–947.

Jakobsen, K. and Hermansen, J.E. (2001). Organic farming – a challenge to nutritionists. Journal of Animal and Feed Sciences 10: Suppl. 1, 29–42.

Kelly, H.R.C., Browning, H.M., Martins, A.P., Pearce, G.P., Stopes, C. and Edwards, S.A. (2005a). The effect of breed on sow performance in organic systems. Proceedings of Congress on Organic Farming, Food Quality and Human Health, Newcastle, UK, 6–9 January. Abstracted in QLIF News May 2005, p. 6; available at http://www.qlif.org/qlifnews/april05/con4.html (accessed November 2015).

Kelly, H.R.C., Browning, H.M., Martins, A.P., Pearce, G.P., Stopes, C., Leifert, C. and Edwards, S.A. (2005b). The effect of breed and husbandry system on performance and of meat and meat products from organic pigs. Proceedings of Congress on Organic Farming, Food Quality and Human Health, Newcastle,

UK, 6–9 January. Abstracted in QLIF News May 2005, p.5; available at http://www.qlif.org/qlifnews/april05/con4.html (accessed November 2015).

NAS–NRC (1998) Nutrient Requirements of Swine, 10th revised edn. National Research Council. National Academy Press, Washington, DC, USA.

Rostagno, M. (2011). Food Safety Fact Sheet, Organic Pork and Food Safety. USDA-ARS-MWA Livestock Behavior Research Unit, Purdue University, West Lafayette, IN 4790, USA.

Saltalamacchia, F., Vincenti, F., Tripaldi, C. and Casa, G.D. (2004) Organic pig breeding. Breeds, feeding, hygiene aspects and type of housing. Rivista di Suinicoltura 45: 75–87.

Uremovic, M., Uremovic, Z., Lukovic, Z. and Konjacic, M. (2003). The influence of genotype and production conditions on the fertility of sows in outdoor system. Agriculturae Conspectus Scientificus (Poljoprivredna Znanstvena Smotra) 68: 245–248.

Wlcek, S. and Zollitsch, W. (2004). Sustainable pig nutrition in organic farming: by-products from food processing as a feed resource. Renewable Agriculture and Food Systems 19: 159–167.

CHAPTER 4

Feeding Organic Poultry

ORGANIC FEEDSTUFFS FOR POULTRY

As with pigs and other livestock, the standards of organic poultry farming are based on the principles of enhancement and utilization of the natural biological cycles in soils, crops and livestock. Accordingly, organic poultry production should maintain or improve the natural resources of the farm system, including soil and water quality.

Producers must keep poultry and manage their waste in a way that supports instinctive, natural living conditions of the birds, yet does not contribute to contamination of soil or water with excessive nutrients, heavy metals, or pathogenic organisms, and optimizes nutrient recycling. Living conditions must accommodate the health and natural behaviour of the birds, providing access to shade, shelter, exercise areas, fresh air, and direct sunlight suitable to their stage of production or environmental conditions, while complying with the other organic production regulations.

Organic poultry production differs from organic livestock production in that the parent stock is not required to be organic.

Feed, including pasture and forage, must be produced organically and health care treatments must fall within the range of accepted organic practices. Organic poultry health and performance should be optimized by application of the basic principles of husbandry, such as selection of appropriate breeds and strains, appropriate management practices and nutrition, and avoidance of overstocking. Rather than being designed to maximize performance, the feeding programmes should be designed to minimize metabolic and physiological disorders, hence the requirement for some forage in the diet. Grazing management should be designed to minimize pasture contamination with parasitic larvae. Housing conditions should be such that disease risk is minimized.

Nearly all synthetic animal drugs used to control parasites, prevent disease, promote growth or act as feed additives in amounts above those needed for adequate growth and health are prohibited in organic poultry production. Dietary supplements containing animal by-products such as meat meal are also prohibited. No hormones can be used, a requirement that is easy to apply in organic poultry production since hormone addition to feed has never been a commercial practice since diethyl stilboestrol (DES), which was used in implantable form in poultry many years ago, was banned in 1959.

Permitted feedstuffs for organic poultry production are listed in Chapter 2, Table

2.1. These feed ingredients are considered to be necessary or essential in maintaining bird health, welfare and vitality; contribute to an appropriate diet fulfilling the physiological and behavioural needs of the species concerned; and do not contain genetically engineered/modified organisms and products thereof.

The specific criteria for feedstuffs and nutritional elements state that:

a) Feedstuffs of plant origin from non-organic sources can only be used under specified conditions and provided they have been produced or prepared without the use of chemical solvents or chemical treatment;
b) Feedstuffs of mineral origin, trace elements, vitamins, and provitamins can only be used if they are of natural origin. In case of a shortage of these substances, or in exceptional circumstances, chemically well-defined equivalent substances may be used;
c) Feedstuffs of animal origin, with the exception of milk and milk products, fish, other marine animals and products derived therefrom, should generally not be used or, as permitted by national legislation.
d) Synthetic nitrogen or non-protein nitrogen compounds shall not be used.

Specific criteria for additives and processing aids state that:

a) Binders, anti-caking agents, emulsifiers, stabilizers, thickeners, surfactants, coagulants: only natural sources are allowed;
b) Antioxidants: only natural sources are allowed;
c) Preservatives: only natural acids are allowed;
d) Colouring agents (including pigments), flavours and appetite stimulants: only natural sources are allowed;
e) Probiotics, enzymes and certain microorganisms are allowed.

The regulations include the requirement for roughage and fresh/dried fodder or silage in the daily ration. All feed ingredients used must be certified as being produced, handled and processed in accordance with the standards specified by the certifying body.

The nutritional characteristics, average composition and (where possible) the recommended inclusion rates (Appendix Table 4.1) of the feedstuffs that are considered most likely to be used in organic poultry diets are set out below. More detailed information is available in the book Nutrition and Feeding of Organic Poultry (Blair, 2008).

Due to a lack of data on feedstuffs that have been grown organically, the nutritional data refer mainly to feedstuffs which have been grown conventionally. Eventually a database of organic feedstuffs composition should be developed.

CEREAL GRAINS AND BY-PRODUCTS

As with pigs, the primary sources of energy in poultry diets are cereal grains. One difference is that poultry have evolved as seed-eaters, so growing and mature birds have an ability to use whole grains whereas pigs require the grains to be processed by grinding or rolling, etc., before being included in feed mixtures.

In addition to the whole grains, the processing of cereals for the human market yields by-products that are important as feed ingredients.

Cereal grains are high in carbohydrate

and they are generally palatable and well digested by poultry. Nutrient variability is usually higher in organic grains than in conventional grains, because of the fertilizer practices in organic grain production, but comparative data are inadequate at present.

Organic cereals tend to have a slightly lower protein content than conventionally grown cereals (Blair, 2012) and cereal by-products tend to be more variable than the grains. As a result, it is advised that periodic checks on the nutrient content of feed mixtures be carried out, to ensure consistency and nutritional adequacy for poultry.

The fibre in grains is contained mainly in the hull (husk) and can be variable, depending on the growing and harvesting conditions. This can affect the starch content of the seed and, as a consequence, the energy value for poultry (Table 4.1). The hull is quite resistant to digestion by poultry and also has a lowering effect on the digestibility of nutrients.

The protein is low in important amino acids such as lysine and methionine, in relation to the bird's requirement.

Grains also tend to be low in vitamins and minerals. Therefore, cereal grain-based diets must be supplemented with other ingredients to meet amino acid and micronutrient requirements.

Feed grains can only meet part of the requirement for dietary nutrients, and other feed components are needed to balance the diet completely. Combining grains and other ingredients into a final dietary mixture to meet the bird's nutritional needs requires information about the nutrient content of each feedstuff and its suitability as a feed ingredient.

Maize, wheat, oats, barley and sorghum are the principal cereals used in poultry feeding, as whole grains and/or grain by-products. The predominant cereal grain used in poultry feeds worldwide is maize.

This is mainly because its energy source is starch, which is highly digestible for poultry. In addition, it is very palatable, is a high-density source of readily available energy, and is free of anti-nutritional factors. The energy value of maize is generally considered the standard with which other energy sources are compared.

The other energy source that meets most of the same criteria as maize is low-tannin sorghum.

Generally, maize and wheat are highest in energy value for poultry, with sorghum, barley, oats and rye being lower. Some rye is used in poultry feeding. Although it is similar to wheat in composition, it is less palatable and may contain ergot, a toxic fungus. Triticale, a hybrid of rye and wheat, is also being used for poultry feeding in some countries.

There do not appear to be any GM (genetically modified) varieties of wheat, sorghum, barley or oats being grown commercially, unlike the situation with maize. In the USA, for instance, substantial quantities of GM maize varieties developed with insect and herbicide resistance are being grown. These modified varieties are not approved for organic poultry production.

Barley

Barley is considered a medium-energy grain, and is an important feedstuff for poultry. The proportion of hull to kernel is variable, resulting in a variable energy value. The nutritive value and suitability of this grain as a feedstuff for growing poultry are influenced also by varying concentrations (commonly 40–150g/kg) of a non-starch polysaccharide (NSP), β-glucan. This compound increases the viscosity of the intestinal contents, resulting in a reduction in nutrient utilisation. NSPs also cause wet and sticky droppings, which may result in foot and leg problems and breast blisters. The negative effects of NSPs can be reduced or eliminated by the addition of β-glucanase enzyme of microbial origin to the feed mixture.

The protein content is higher than in maize and can range from about 90 to 160g/kg. The amino acid profile is better than in maize and is closer to that of oats or wheat and the bioavailability of the essential amino acids (EAA) is high.

When formulating barley-based feed mixtures the low content of linoleic acid should be taken into account, otherwise egg size may be reduced.

Hull-less barley varieties have been developed in which the hull separates during threshing. These varieties contain more protein and less fibre than conventional barley, and theoretically should be superior in nutritional value to conventional barley. However, several studies have failed to show improved growth performance over conventional barley in poultry diets. As a result, hull-less barley should be used as a substitute for conventional barley on the assumption that they are similar in nutritive value.

Barley should be ground to a fine, uniform consistency for mixing into diets to be pelleted or crumbled. A coarser grind can be used for mash diets.

Oats

Oats can be grown on many organic farms and utilized on-farm. Oats are higher in hull, fibre and ash contents and are lower in starch than maize, grain sorghum or wheat. The proportion of hulls can range from 200 to 450g/kg. Thus oats have a much lower energy content for poultry than other main cereals. The nutritional value is inversely related to the hull content, which can be approximated from the thousand kernel weight.

Inclusion of oats (and barley) in the diet adds fibre and increases the bulk density of feed. The added fibre has a beneficial effect on gut health and function, but is more difficult to digest than starch unless supplementary enzymes are included in the feed. The result of high levels of oats (or barley) in the diet is wet litter due to the digesta in the gut being very viscous. Supplementation with glucanase enzyme avoids these adverse effects.

Oats vary in protein content from about 110 to 170g/kg, with an amino acid profile similar to wheat, being limiting in lysine, methionine and threonine. The protein content is affected mainly by the proportion of hull, since it is present almost entirely in the kernel. Oats have a higher oil content than maize, but this does not compensate for the high fibre content. Owing to their high fibre and low energy content, oats are best used in pullet developer diets and in diets for breeder flocks. Newer varieties of oats with a lower hull content may be more suitable in diets for growing meat-type birds.

Maize

This cereal is also known as corn or Indian corn in the Americas. It can be grown in more countries than any other grain crop because of its versatility. It is the most important feed grain in North America because of its palatability, high energy value, and high yields of digestible nutrients per unit of land. Consequently, it is used as a yardstick in comparing other feed grains for poultry feeding. The plains of the USA provide some of the best growing conditions for maize, making it the world's top maize producer.

Maize is high in carbohydrate, most of which is highly digestible starch, and it is low in fibre. It has a relatively high oil content, and thus a high energy value for poultry. Other grains, except wheat, have a lower energy value than maize. Maize oil has a high proportion of polyunsaturated fatty acids and is an excellent source of linoleic acid.

The use of yellow maize grain should be restricted in poultry diets if it results in carcass fat that is too soft or too yellow. White maize can be used to avoid the fat colouration. Yellow and white maize are comparable in energy, protein, and minerals. Yellow maize is a valuable component of layer diets when yellow-pigmented yolks are desired.

Protein concentration in maize is normally about 85g/kg but the protein is not well balanced in amino acid content, with lysine, threonine, isoleucine and tryptophan being limiting. Maize is very low in calcium (about 0.3g/kg). It contains a higher level of phosphorus (2.5–3.0g/kg) but much of the phosphorus is bound in phytate form that it is poorly available to poultry. As a result, a high proportion of the phosphorus passes through the gut and is excreted in the droppings. The diet can be supplemented with phytase enzyme to improve phosphorus utilization. Another approach is to use one of the newer low phytic acid maize varieties that have about 35% of their phosphorus bound in phytate acid compared with 70% for conventional maize. These varieties allow the phosphorus to be more effectively utilized by the bird and less excreted in the droppings. As with other improved varieties of maize, producers wishing to use such varieties should check their acceptability with the organic certifying agency.

The quality of maize is excellent when harvested and stored under appropriate conditions, including proper drying to 10–12% moisture. Fungal toxins (zearalenone, aflatoxin and ochratoxin) can develop in grain that is harvested damp or allowed to become damp during storage. These toxins have adverse effects in poultry.

Maize is suitable for feeding to all classes of poultry, with no feed inclusion limitations. One of the benefits of using yellow maize in poultry diets is that birds are attracted to the yellow colour of the grain. Maize should be medium ground to a uniform particle size, slightly smaller for chicks and larger for adult poultry when included in mash. A fine grind is recommended for inclusion in pelleted diets. The grain should be mixed into feed immediately after grinding since it is likely to become rancid during storage.

Sorghum

Sorghum is one of the most drought-tolerant cereal crops and is more suited than maize to harsh weather conditions such as high temperature and less consistent moisture. It is generally higher in crude protein than maize and is similar in energy content. However, one disadvantage of grain sorghum is that it can be more variable in composition because of growing conditions. Crude protein content usually averages around 89g per kg, but can vary from 70 to 130g per kg. Therefore, a protein analysis prior to formulation of diets is recommended.

The hybrid yellow-endosperm varieties are more palatable to poultry than the darker brown sorghums, which possess a higher tannin content designed to deter wild birds from damaging the crop. High-tannin cultivars can be used successfully after storage under high-moisture conditions (reconstitution) for a period of 10 days to inactivate the tannins. Reconstitution has been shown to result in an increased protein digestibility and utilisation of energy, indicating a way for poultry producers with only high-tannin sorghum available to use this grain in poultry diets.

Research results show that low-tannin grain sorghum cultivars can be used successfully as the main or only grain source in poultry diets. However, proper grinding of grain sorghum is important to break the hard seed coat.

Wheat

Wheat is commonly used for feeding to poultry when it is surplus to human requirements or is considered not suitable for the human food market, otherwise it may be too expensive for poultry feeding. Some wheat is grown, however, for feed purposes. By-products of the flour-milling industry also are very desirable ingredients for poultry diets.

During threshing, the husk – unlike that of barley and oats – detaches from the grain, leaving a less fibrous product. As a result, wheat is almost as high in energy as maize but contains more protein, lysine and tryptophan. Thus, it can be used as a replacement for maize as a high-energy ingredient in poultry diets and it requires less protein supplementation than maize.

Wheat is very palatable if not ground too finely and can be used efficiently by all classes of poultry. Good results are obtained when wheat is coarsely ground (hammer mill screen size of 4.5–6.4mm) for inclusion in mash diets. Finely ground wheat is not desirable because it is too powdery, causing the birds difficulty in eating unless the diet is pelleted. In addition, finely ground wheat absorbs moisture readily from the air and saliva in the feeder, becoming sticky. This can result in feed spoilage and reduced feed intake. Also, feed containing finely ground wheat can bridge and not flow well in feeding equipment.

One benefit of wheat as a feed ingredient for diets to be pelleted is that it improves pellet quality due to its gluten content. As a result, the use of a pellet binder may be unnecessary.

The amount of wheat in conventional poultry diets is generally limited to 300g/kg inclusion unless supplemented with xylanase enzyme to improve digestion. These enzymes are permitted in organic production.

One concern about wheat is that the energy and protein contents are more variable than in other cereal grains such as maize, sorghum and barley, therefore periodic testing of batches of wheat for nutrient content is necessary. Evidence of the variability in the nutrient content of 16 cultivars of wheat was obtained in a Danish study (Steenfeldt et al., 2001). Protein content ranged from 112 to 127g/kg dry matter, starch content from 658 to 722g/kg dry matter and NSPs from 98 to 117g/kg dry matter. When fed to growing chickens the variability in chemical composition was expressed in differences in growth performance and in intestinal effects, especially when wheat was included at high levels in the diet. Dietary apparent metabolizable energy (AME) levels ranged from 12.66 to 14.70MJ/kg dry matter with a dietary wheat content of 815g/kg and from 13.20 to 14.45MJ/kg dry matter with a dietary wheat content of 650g/kg. Milling quality wheat resulted in better performance than feed grade wheat.

In spite of the recorded variability in nutrient content, there is a large body of evidence indicating that wheat can be used effectively in diets for growing and laying poultry.

Replacing maize with wheat reduces the total xanthophyll content of the feed, reducing the yellow pigmentation of the skin, shanks and egg yolk. Therefore, supplementary sources of xanthophylls may have to be used in grower and layer diets when wheat is used to replace maize.

Wheat Middlings

Milling by-products from flour production can also be used in poultry feeding. These by-products are usually classified according to their protein and fibre contents and are traded under a variety of names such as pollards, offals, shorts, wheatfeed and midds (wheat middlings). In general those by-products with low levels of fibre are of higher nutritive value for poultry. The composition and quality of middlings vary greatly due to the proportions of fractions included, also the amount of screenings added and the fineness of grind.

Wheat middlings are used commonly in poultry diets, as a partial replacement for grain and protein supplement. This by-product is included commonly in commercial feeds as a source of nutrients and because of its beneficial influence on pellet quality. When middlings (or whole wheat) are included in pelleted feeds, the pellets are more cohesive and there is less breakage and fewer fines.

Spelt

Spelt is a subspecies of wheat grown widely in Central Europe. Expansion of this crop is likely in Europe because of the current shortage of high-protein organic feedstuffs. This crop appears to be generally more winter hardy than soft red winter wheat, but less winter hardy than hard red winter wheat.

A report cited by Blair (2008) found an average protein value of 166g per kg in spelt varieties in the US vs 134g per kg in wheat varieties. The average lysine content was 29.3g per kg of protein in spelt vs 32.1g per kg of protein in wheat. The nutrient profile of both grains was found to be greatly influenced by cultivar and location.

Triticale

Triticale is a hybrid of wheat and rye, developed with the aim of combining the grain quality, productivity and disease resistance of wheat with the vigour, hardiness, and high lysine content of rye. This grain, like wheat, varies in nutrient composition and can be used successfully in poultry diets as a partial or complete replacement for wheat. As with wheat, egg yolk colour can be reduced greatly when the diet contains high levels of triticale but this should not occur when the hens have access to good quality forage.

Rye

Rye has an energy value intermediate to that of wheat and barley, and the protein content is similar to that of barley and oats. However, its nutritional value is reduced by the presence of several anti-nutritional fractions such as β-glucans and arabinoxylans, which are known to cause increased viscosity of the intestinal contents and reduced digestibility, and other undesirable effects such as an increased incidence of dirty eggs. These effects are more pronounced in hot and dry environments that accentuate the rate of cereal ripening prior to harvest, as occurs in Spain and other Mediterranean countries. Rye may also contain ergot, which is a toxic fungus that reduces poultry health and performance.

SUPPLEMENTARY PROTEIN SOURCES

The main protein sources used in organic poultry feeding are soya beans, canola and groundnuts (peanuts). These crops are grown primarily for their seeds, which produce oils for human consumption and

Component g/kg unless stated	Barley	Wheat	Wheat Middlings	Oats	Maize	Sorghum	Triticale	Rye
Dry matter	890	880	890	890	890	880	900	880
ME kcal/kg	2640	3160	2200	2610	3350	3290	3160	2630
ME MJ/kg	11.5	12.9	9.0	10.9	14.2	13.7	12.7	11
Crude protein**	113	135	159	115	83	92	125	118
Lysine	4.1	3.4	5.7	4	2.6	2.2	3.9	3.8
Methionine + Cystine	4.8	4.9	5.8	5.8	3.9	3.4	4.6	3.6
Crude fibre	50	26	73	108	26	27	21.6	22
Crude fat	19	20	42	47	39	29	18	16
Linoleic acid	8.8	9.3	17.4	26.8	19.2	13.5	7.1	7.6
Calcium	0.6	0.6	1.2	0.7	0.3	0.3	0.7	0.6
Phosphorus	3.5	3.7	9.3	3.1	2.8	2.9	3.3	3.3
Non-phytate (available) phosphorus	1.7	1.8	3.8	0.7	0.4	0.6	1.4	1.1

TABLE 4.1 Average composition of common cereal grains and grain products for organic poultry feeding.*

* Air-dry basis. ** N × 6.25

industrial uses. Cottonseed is used in poultry feeding in some regions, also sunflowers, peas and lupin seed.

Most oilseeds and legumes contain anti-nutritive factors, which are mainly destroyed by moderate heating. Organic poultry producers who are able to grow protein crops on-farm have, therefore, to heat process the seed in a suitable way before including it in feed mixtures.

Developments are taking place that allow oilseed crops to be used on-farm for poultry feeding, providing high-energy, high-protein feed ingredients. These developments, where warranted technically and economically, are of great relevance to organic animal production.

New cultivars of some oilseeds and legumes have been developed that are naturally low in anti-nutritive factors, permitting higher levels of the unprocessed seed to be included in poultry diets without ill effect.

Only those meals obtained by mechanical extraction of the oil from the seed are acceptable for organic diets.

As a group, the oilseed meals are high in protein content but are low in important amino acids such as lysine and methionine. The extent of dehulling affects the protein and fibre contents, whereas the efficiency of oil extraction influences the oil content and thus the energy content of the meal.

Oilseed meals are generally low in calcium and high in phosphorus content, although much of the phosphorus is present as phytate, which has a low biological availability for poultry. The non-phytate phosphorus content is therefore a more useful measure in evaluating phosphorus contents of feedstuffs for poultry.

Soya Beans and Soya Bean Products

Soya beans and soya bean meal are now used widely in animal feeding. Several genetically modified strains of soya beans are being grown, therefore organic producers have to be careful to select non-GM soya feedstuffs. The major GM crops grown in North America are soya beans, maize, canola and cotton and these are exported to many countries.

Soya beans are grown as a source of protein and oil for the human market and for the animal feed market. Soya bean meal is generally regarded as the best plant protein source in terms of its nutritional value. Also, it has a complementary relationship with cereal grains in meeting the amino acid requirements of poultry. Consequently, it is the standard to which other plant protein sources are compared.

Whole soya beans contain 360 to 370g per kg protein, whereas soya bean meal contains 410 to 500g per kg depending on efficiency of the oil extraction process and the amount of residual hulls present. Relative to other oilseed meals, soya bean protein has a good balance of essential amino acids for poultry. The content of lysine in soya bean protein is exceeded only in pea, fish, and milk proteins. The amino acid availability in soya bean meal is higher than in other oilseed meals and the energy content is also substantially higher than in other oilseed meals.

Soya beans contain 150 to 210g per kg oil, most of which is removed in the oil extraction process. The oil has a high content of the polyunsaturated fatty acids, linoleic and linolenic acids.

Only the meal remaining after the mechanical process of oil removal is acceptable in organic diets. An advantage of this (expeller) process over the conventional solvent extraction process is that it is less efficient in extracting the oil, and the higher content of oil gives the meal a higher energy content. A disadvantage is that that the heat generated by friction during the expeller

process, while inactivating anti-nutritional factors present in raw soya beans, subjects the product to a higher processing temperature than in the solvent extraction process, and makes the protein harder to digest. Expeller soya bean meal is thus favoured for dairy cow feeding since the higher content of less available (so-called 'rumen by-pass') protein results in improved milk production. Consequently, most of the expeller soya bean meal available in North America is used by the dairy feed industry, and may be difficult to obtain by the organic poultry producer.

More recently a new process known as extruding-expelling, has been developed. Extruders are machines in which soya beans or other oilseeds are forced through a tapered die. The frictional pressure causes heating. In the extruding-expelling process a dry extruder in front of the screw presses eliminates the need for steam. These plants are relatively small, typically processing 5 to 25 tonnes of soya beans per day. The dry extrusion-expelling procedure results in a meal with greater oil content than conventional solvent extracted meal, but with similar low trypsin inhibitor values. The nutritional characteristics of extruded-expelled meal have been shown to be similar to those of mechanically processed meal, therefore this process should be of interest to organic poultry producers since the soya product qualifies for acceptance in organic diets.

Yet another process being used in small plants is extrusion, but without removal of the oil, the product being a full-fat meal. Often these plants are operated by co-operatives and should be of interest to organic poultry producers, since the product also qualifies for acceptance in organic diets.

Soya bean meal has an excellent amino acid profile for poultry. This feature allows the formulation of diets that contain lower total protein than with other oilseed meals, thereby reducing the amount of nitrogen to be excreted by the bird and reducing the nitrogen load on the environment.

Soya bean meal is generally low in minerals and vitamins. About two-thirds of the phosphorus in soya beans is bound as phytate and is mostly unavailable to poultry. This compound also binds with mineral elements including calcium, magnesium, potassium, iron, and zinc, rendering them unavailable to poultry. Therefore, it is important that diets based on soya bean meal contain adequate amounts of essential minerals. Another approach to the phytate problem is to add phytase, a phytic acid degrading enzyme, to the feed to release phytin-bound phosphorus. A benefit of this approach is that less phosphorus needs to be added to the diet, reducing excess phosphorus loading into the environment.

Use of Whole Soya Beans

Since soya beans contain a high quality protein and are a rich source of oil they have the ability to provide major amounts of protein and energy in the diet. Use of full-fat beans is a good way of increasing the energy level of the diet, particularly when they are combined with low-energy ingredients. Also, this is an easier way to blend fat into a diet than by the addition of liquid fat. Several reports show that properly processed whole soya beans (to inactivate the anti-nutritional factors) can be used effectively in poultry diets as a partial or complete replacement for soya bean or other protein meals.

Extruded soya beans are whole soya beans that have been exposed to a dry or wet (steam) friction heat treatment without removing any of the component parts. Roasted soya beans are whole soya beans that have been exposed to a heat or micronized treatment without removing any of the component parts.

Feeding whole soya beans may have adverse effects on the carcass fat of meat-type poultry. A standard recommendation is that the soya bean product should be limited to providing a dietary addition of 20g soya bean oil per kg diet, in order to ensure an acceptable carcass fat quality and good pellet quality. This generally limits the incorporation rate of soya beans in the feed of market birds to 100g per kg.

Including full-fat soya beans in poultry diets is reported to reduce aerial dust levels and is therefore likely to benefit the health of animals and workers in buildings. Owing to possible rancidity problems, diets based on full-fat soya beans should be used immediately and not stored. Otherwise, an approved antioxidant should be added to the dietary mixture.

Canola (Improved Rapeseed)

Canola ranks fifth in world production of oilseed crops, after soya beans, sunflowers, peanuts and cottonseed. It is the primary oil seed crop produced in Canada. This crop is widely adapted but appears to grow best in temperate climates, being prone to heat stress in very hot weather. As a result, canola is often a good alternate oilseed crop to soya beans in regions not suited for growing soya beans.

Canola seed contains about 400g oil, 230g protein, and 70g per kg crude fibre. The oil is high in polyunsaturated fatty acids (oleic, linoleic, and linolenic), which makes it valuable for the human food market. It can also be used in animal feed. The oil is, however, highly unstable due to its content of polyunsaturated fatty acids and, like soya bean oil, can result in soft body fat. For organic feed use the extraction has to be done by mechanical methods such as crushing (expeller processing), avoiding the chemical solvent used in conventional processing.

Two features of expeller processing are important. As with soya bean meal, the amount of residual oil in the meal varies with the efficiency of the crushing process, resulting in a product with a more variable energy content than the commercial, solvent-extracted product. Therefore, more frequent analysis of expeller canola meal for oil and protein contents is recommended and more conservative limits should be placed on the levels of expeller canola meal used in poultry diets.

The extracted meal is a high quality, high-protein feed ingredient, with about 350–400g per kg protein depending on type. The lysine content of canola meal is lower and the methionine content is higher than in soya bean meal. Otherwise it has a comparable amino acid profile to soya bean meal.

Owing to its high fibre content (>110g per kg), canola meal contains about 15 to 25% less energy for poultry than soya bean meal. Dehulling can be used to increase the energy content. Like other plant protein meals, canola meal is high in phytate, which reduces the availability of many mineral elements unless the diet is supplemented with a phytase enzyme.

It is been shown that brown-shelled layers based on Rhode Island Red stock produce 'fishy eggs' when fed diets containing canola meal. This is due to high levels of choline (a vitamin) and sinapine (phenolic) compounds in canola that are converted to trimethylamine in the gut. This compound has a 'fishy' odour. White-shelled layers have the ability to break down the trimethylamine in the gut to the odourless oxide that does not result in the problem. However, certain brown-shelled layers do not produce enough of the trimethylamine oxidase enzyme to break down the trimethylamine and instead deposit the compound in the egg. Consequently, canola should not be fed to brown-shelled laying stock

of Rhode Island parentage, or only at low levels. In North America, an upper limit of 30g/kg canola meal is recommended in diets for brown-shelled layers. Higher levels may be used in countries in which high levels of fishmeal have been historically used in feeds, the consumer being accustomed to eggs with a fishy flavour.

Canola meal can be used in all types of poultry feed. Owing to its relatively low energy value for poultry, it is best used in layer and breeder diets rather than in high-energy grower feeds. The meal has been included successfully in a range of poultry diets at levels up to 200g/kg.

Full-Fat Canola (Canola Seed)

A more recent approach with canola is to include the unextracted seed in poultry diets, as a convenient way of providing both supplementary protein and energy. Good results have been achieved with this feedstuff, especially with the low-glucosinolate cultivars. However, there are two potential problems that need to be addressed, as with full-fat soya beans. The maximum nutritive value of full-fat rapeseed is only obtained when the seed is mechanically processed and heat-treated to allow glucosinolate destruction and release of the oil.

Once ground, the oil in full-fat canola becomes highly susceptible to oxidation, resulting in undesirable odours and flavours. The seed contains a high level of α-tocopherol (vitamin E), a natural antioxidant, but additional supplementation with an acceptable antioxidant is needed if the ground product is to be stored. A practical approach to the rancidity problem is to grind just sufficient canola for immediate use.

Several studies have shown that canola seed can be included in layer, broiler and turkey diets but that at high inclusion levels the performance results were lower than expected unless the diets were steam pelleted. One interesting finding for organic producers was that there was a linear increase in contents of polyunsaturated fatty acids in eggs yolks with increasing content of canola seed in the diet. Also, the fat in skeletal muscle, skin, sub-dermal fat, and abdominal fat of birds fed diets containing canola seed had the highest contents of polyunsaturated fatty acids.

On the basis of these and related findings it is suggested that the level of full-fat canola in poultry diets be limited to 50 to 100g per kg and that the seed be subjected to some form of heat treatment either before or during feed preparation.

Faba (Field) Beans

Beans are well established as a feedstuff for horses and ruminants and this crop is now receiving more attention as a feedstuff for poultry, particularly in Europe, because of the deficit in protein production there. In 2014 the EU-28 feed industry used more than 52 million tonnes of protein feeds but produced only 29 million tonnes, the balance being imported. The most feasible expansion in protein feedstuffs production in EU countries is likely to be from crops of the legume family (such as beans, peas, lupins and soya beans).

Field beans contain about 240–300g/kg protein, the protein being high in lysine and (like most legume seeds) low in methionine and cystine. The low content of methionine in field beans (and in some other supplementary protein sources) places a limitation on the amount that can be included in poultry diets. Methionine is a major component of feathers and is very important for proper feather formation. A deficiency of methionine results in poor feather growth and increased feather pecking, which can

lead to cannibalistic behaviour. A methionine deficiency also results in poor feed conversion, retarded growth in meat birds, and reduced egg production in layers and breeders.

The difficulty in meeting the methionine requirements of organic poultry has resulted in the organic standards in several countries having to be relaxed temporarily to allow the use of synthetic methionine. The exemption was introduced to allow the organic industry to find alternative sources of methionine. This problem has not yet been resolved and in the USA, for instance, the National Organic Standards Board recommended that the use of synthetic methionine be continued after October 1, 2012, but in reduced amounts. It is likely that this temporary exemption will continue in the USA until 2017 or 2021.

The debate on whether pure forms of amino acids such as methionine (and lysine) are acceptable and necessary supplements for some organic diets is likely to continue for some time.

Further research is needed to determine whether poultry can produce meat and eggs effectively when allowed access to pasture and forage as a methionine source. Also, the value of by-products such as non-GM maize gluten meal and potato meal as methionine sources needs to be researched.

At present it is prudent for organic producers to seek advice on whether or not to take advantage of relaxed regulations to include synthetic methionine (and lysine) in their poultry feed formulations, to ensure nutritional adequacy. The use of synthetic methionine in poultry diets makes it possible to formulate to lower levels of dietary protein that still meet the daily methionine requirement. Methionine supplementation is typically in the form of a dry powder or as a liquid.

Examples of feed mixtures with and

without methionine supplementation are shown later in this section.

Some research indicates that meat chickens in the grower and finisher phase can obtain sufficient methionine while foraging pastures. This would include the plant material consumed as well as insects, worms, etc. However, obtaining significant methionine from pasture depends greatly on forage composition and management, as well as environmental conditions.

Some investigations have involved the use of slower growing strains of birds to test whether they have a lower methionine requirement. Recent research in the USA by Fanatico et al. (2009) indicates that slower and faster growing strains have similar methionine requirements.

In general, field beans are of limited value due to their extremely low concentration of methionine (and cystine), but when combined with other feed ingredients that provide sufficient methionine can be used at levels not exceeding 100–150g/kg diet. Small-seeded cultivars with low levels of anti-nutritive factors are preferred. For use in poultry diets, the beans should be ground to pass through a 3mm screen.

Field Peas

Field peas are grown primarily for human consumption, but they are now used widely in poultry feeding. Some producers grow peas in conjunction with barley, as these two ingredients can be successfully incorporated into a poultry feeding programme.

Both green and yellow cultivars are grown, which are similar in nutrient content. Those grown in North America and Europe, both green and yellow, are derived from white-flowered varieties. Brown peas are derived from coloured flower varieties and have higher tannin, lower starch, higher protein and higher fibre contents than green

and yellow peas. These varietal differences account for much of the reported variation in nutrient content. White-flowered cultivars are preferred for poultry feeding. Pea protein concentrate from starch production may also be available as an organic feed ingredient.

Peas are slightly lower in energy content than high-energy grains such as maize and wheat, with about 2,600kcal/kg metabolizable energy, but have a higher protein content (about 230g/kg) than grains. Thus, they are regarded primarily as a protein source. Pea protein is particularly rich in lysine, but relatively deficient in tryptophan, methionine and cystine.

Brown-seeded cultivars are best avoided for poultry feeding since they contain a higher level of anti-nutritional factors such as tannins. Most of the field peas grown in Europe and Canada have a zero tannin content.

Field peas can be used successfully in poultry diets at levels up to 300g/kg when the feed mixture can be formulated to contain a sufficiently high level of energy and essential amino acids. As with faba beans, the low methionine content of peas limits the amount that can be included in organic poultry diets. They should be ground before inclusion in feed mixtures, but production of 'fines' is to be avoided during grinding unless the diets are to be pelleted. Grinding to an extremely small particle size is not economical and may lead to interference with feed intake due to build-up of material in the beak and result in beak necrosis.

Some commercial enzyme supplements containing xylanases and β-glucanases have been reported to reduce the viscosity of intestinal contents and increase protein digestibility when diets containing high levels of peas have been fed to poultry. At the highest levels of recommended use the droppings may be slightly wetter than usual, but this should not be a problem with poultry with access to outdoors.

Research cited by Blair (2008) reported that fig powder acted as a source of enzymes that reduced intestinal viscosity and improved digestibility when diets containing peas were fed to broilers – an approach that may be of interest to organic producers seeking a natural source of these enzymes.

Sunflower Seed and Meal

Sunflower seed is grown mainly for oil production for the human market, leaving the extracted meal available for animal feeding. Sunflower oil is highly valued for its high content of polyunsaturated fatty acids and stability at high temperatures.

Sunflower seed surplus to processing needs and seed unsuitable for oil production may also be available for feed use. On-farm processing of sunflower seed is being done in countries such as Austria.

The protein is relatively deficient in important amino acids, which limits the amount that can be included in organic poultry diets, unless supplements of pure amino acids such as lysine and methionine are acceptable. The crude fibre content of whole (hulled) sunflower meal is around 300g per kg and with a complete decortication (hull removal) the fibre content is around 120g per kg. The high fibre content limits the amount of sunflower meal that can be incorporated into poultry diets. In contrast to other major oil seeds and oilseed meals, sunflower seeds and meals are relatively free of anti-nutritional factors.

Sunflower meal can be used successfully in organic poultry diets to replace up to 50% of the soya bean meal, providing the contents of energy and essential amino acids are adequate. Owing to its high oil content, unextracted sunflower seed provides a convenient method of providing additional energy to poultry diets in the form of sunflower oil.

Lupins

The development of low-alkaloid (sweet) cultivars in Germany in the 1920s allowed lupin seed to be used as animal feed. Prior to that time the seed was unsuitable for animal feeding because of a high content of toxic alkaloids. The shortage of organic protein feedstuffs in Europe has stimulated interest in lupins as an alternative protein source.

In Australia, where much of the recent research has been done, the main species of lupins used in poultry diets are *Lupinus angustifolius*, and *L. luteus*. This research indicated that *L. luteus* (yellow lupin) has significant potential as a poultry feedstuff but that *L. albus* is not recommended for use in poultry diets because it causes a reduction in growth rate. *L. luteus* is native to Portugal, Western Spain and the wetter parts of Morocco and Algeria.

Lupin seed is lower in protein and energy contents than soya bean meal. A factor related to the relatively low energy level is the high content of fibre, 130–150g per kg in *L. luteus*. Another factor is the type of carbohydrate in lupins, being different from that in most legumes with negligible levels of starch and high levels of soluble and insoluble non-starch polysaccharides (NSPs) and oligosaccharides. These compounds influence the speed of feed through the gut, the utilisation of nutrients and also the microflora of the gastrointestinal tract. They also influence water intake and can cause wet droppings. One of the main ways of addressing the adverse effects of these factors and improving the utilization of lupin seed is to add a supplement of appropriate enzymes to the diets.

Dehulling the seed can produce a feedstuff more comparable to soya bean meal.

Australian research suggests an optimal inclusion rate of less than 100g sweet lupin seed per kg diet for growing birds. Birds grown to 21 days on diets containing faba beans, field peas or sweet lupin seed showed better growth rate and feed efficiency than those given chick peas. Steam pelleting of the diet improved growth rate and feed conversion efficiency. Digesta viscosity and excreta stickiness scores were much higher in diets with sweet lupin seed.

German research suggested that up to 300g/kg yellow lupin seed can be included in diets for growing birds, provided amino acid supplementation is used and the energy level is maintained correctly.

The above findings suggest that lupin seed can be a useful alternative to soya beans in poultry diet. However, their inclusion, like that of several other alternative protein sources, will be constrained by the availability of a supplementary source of amino acids. One disadvantage of lupins is a high fibre content that makes them unsuitable for young poultry unless they are dehulled. More mature birds, with their better-developed gut systems, are more able to digest lupins than younger birds.

Groundnuts

Groundnuts (also known as peanuts) are not included as an approved feedstuff in either the EU or New Zealand lists but should be acceptable for organic poultry diets if grown organically.

Groundnuts not suitable for human consumption are used in the production of groundnut (peanut) oil. Raw groundnuts contain 400–550g oil/kg and the by-product of oil extraction, groundnut meal, is widely used as a protein supplement in poultry diets in several countries.

Mechanically extracted meal contains about 50–70g oil/kg, and tends to become rancid during storage in hot weather.

The crude protein content of extracted meal ranges from 410 to 500g/kg and is relatively low in lysine and methionine, which limits the amount that can be included in poultry diets. Groundnut meal is a useful protein source in countries such as Nigeria that are able to grow this crop, and has been shown to be superior to cottonseed cake for use in maize-based diets for growing poultry.

One potential problem with groundnuts is that they are subject to contamination with moulds, such as *Aspergillus flavus,* which produces aflatoxin that is toxic to farm stock and humans. Periodic testing for mould contamination is recommended.

Both groundnuts and groundnut meal can be used successfully in poultry diets. Research findings indicate that productivity of growing and laying birds is reduced when more than 50% of the protein from soya bean meal is replaced by groundnut meal, unless supplemented with protein sources rich in lysine and methionine. Some research has reported that eggs had better interior egg quality when the hens were fed groundnut meal.

Cottonseed Meal

Most of the cottonseed meal is used in ruminant diets, but it can be used in poultry diets when its limitations are taken into account in feed formulation. The fibre content is higher in cottonseed meal than in soya bean meal, and the energy value is inversely related to the fibre content. Expeller cottonseed cake is used extensively in cotton-producing countries such as Pakistan as a protein supplement in poultry diets.

The inclusion of cottonseed meal in poultry diets is further limited by its amino acid profile (low in lysine) and by the presence of gossypol, a compound found in the pigment glands of the seed. The gossypol problem does not occur with glandless cultivars of cottonseed. Gossypol causes yolk discolouration in laying hens, particularly after storage of the eggs. In addition, the fatty acids in cottonseed oil result in a pinkish colour in the egg albumen. Increased mortality due to heart failure has also been reported in poultry, but only when the free gossypol content of the diet approached 400mg per kg.

A general recommendation is that cottonseed meal can replace no more than 50% of the soya bean meal or other protein supplement in the diet. Supplementation of the diet with iron salts, such as ferrous sulphate, has been shown to be effective in blocking the toxic effect of dietary gossypol when gossypol-free cottonseed meal is not available. However, it is unlikely that this treatment would be acceptable for organic poultry diets.

The available findings suggest that, in general, cottonseed meal can be used at low levels in organic diets for growing poultry and that it is utilized most effectively when its lower digestibility and possible gossypol content are taken into account in feed formulation. A general recommendation is that the cottonseed meal used in organic poultry feeding should be from a glandless cotton cultivar and devoid of gossypol.

Fishmeal

Though not an organic product in the strict sense, fishmeal is approved for use in organic poultry diets. A requirement for use in organic production, is that the fishmeal must be derived from sustainable fish stocks and have been oil-extracted by a mechanical method.

Fishmeal is known to stimulate feed intake in birds, attributed to poultry having an instinctive craving for meat proteins (and possibly the salty taste).

Fishmeal used for feed in North America must not contain more than 100g moisture or 70g salt per kg, and the amount of salt must be specified if it is greater than 30g per kg. Antioxidants are commonly added to the meal to prevent oxidation and spoilage, therefore this aspect should be checked for acceptability in organic diets.

Fishmeal is one of the best protein sources for poultry, typically 500 to 750g/kg and essential amino acids, particularly lysine, that are deficient in many cereal grains and other feedstuffs. It is also a rich source of vitamins and minerals. Owing to its high price it is generally used to balance the amino acid contents of diets rather than as a major source of protein. It is a main (often the only) supplementary protein source in some Asian countries.

In general, fishmeal can be included at up to 80g/kg in diets for young birds, and less than 40g/kg for older meat birds and layers. Higher levels must be avoided in finishing and laying diets, unless a fishy taint in meat and eggs is acceptable.

TUBER ROOTS, THEIR PRODUCTS AND BY-PRODUCTS

This group includes sugar beet pulp, dried beet, potatoes, sweet potatoes and manioc as roots and tapioca.

Cassava

Cassava is an approved ingredient in organic poultry diets, although in many countries it will represent an imported product not produced regionally. Fresh cassava contains about 650g/kg moisture. The dry matter is high in starch but very low in protein (20

Component g/kg unless stated	Soya beans	Soya bean meal	Canola seed	Canola meal	Groundnut meal	Faba beans	Field peas	Sunflower meal dehulled	Lupin seed white
Dry matter	900	900	940	940	930	870	890	930	890
ME kcal/kg	3850	2750	4640	2500	2300	2430	2570	2350	2950
ME MJ/kg	16.1	11.5	17.4	10.5	9.6	10.3	10.7	9.8	12.2
Crude protein**	352	420	242	352	432	254	228	414	349
Lysine	22.2	27.0	14.4	20	14.8	16.2	15	16.1	15.4
Methionine + Cystine	10.8	12.2	8.6	16	11	5.2	5.2	16.3	7.8
Crude fibre	43	65	74	120	69	73	61	122	110
Crude fat	180	81	400	107	65	14	12	80	97.5
Linoleic acid	104	27.9	110	12.8	17.3	5.8	4.7	19	35.6
Calcium	2.5	2.0	3.9	6.8	1.7	1.1	1.1	3.9	3.4
Phosphorus	5.9	6.1	6.4	11.5	5.9	4.8	3.9	10.6	3.8
Non-phytate (available Phosphorus)	2.3	2.3	2.0	4.0	2.7	1.9	1.7	2.7	1.6

TABLE 4.2 Average composition of common protein feedstuffs for organic poultry feeding.*

* As-fed basis. ** N × 6.25

to 30g/kg, of which only about 50% is in the form of true protein).

Cassava meal is an excellent energy source in poultry diets because of its highly digestible carbohydrates (700–800g per kg), mainly in the form of starch. However, its main drawback is the negligible content of protein and micronutrients, including carotenoids for egg and skin pigmentation.

Fresh cassava needs to be processed by boiling, roasting, soaking, ensiling, or sun drying to reduce the levels of cyanogenic glucosides (mainly linamarin), which on ingestion are hydrolyzed to hydrocyanic acid, a poisonous compound. Sulphur is required by the body to detoxify the cyanide, therefore the diet needs to be adequate in sulphur-containing compounds such as methionine and cystine. The normal range of cyanide in fresh cassava is about 15 to 400mg/kg fresh weight. A dietary cyanide content in excess of 100mg/kg diet has been shown to affect growing birds adversely but laying hens can be affected by levels as low as 25mg total cyanide/kg diet.

Research studies indicate that low-cyanide cassava root meal can be included successfully in balanced poultry diets at levels up to 250g/kg as a cereal replacement. It is unlikely that organic diets can be formulated to contain such high levels due to difficulties in maintaining the correct concentration of supplementary protein and amino acids. In addition, egg yolks from hens fed diets containing high levels of cassava are very pale in colour and not acceptable in many markets. In this situation the diet requires supplementation with a source of carotenoids (such as maize or marigold extract) or the hens have to be given access to green crops that contain yolk-pigmenting compounds.

The dustiness of cassava-based diets can be reduced by adding molasses or oils to the feed mixture, or by pelleting.

Potatoes

Potatoes have potential as a poultry feed but are more widely used in pig and ruminant feeding because they are more difficult to incorporate into poultry feed mixtures than other feed ingredients. Nevertheless, cooked potatoes were used commonly as poultry feed in Europe during and after the Second World War when feedstuffs were in short supply.

As with most root crops, the major drawback is the relatively high content of water and consequent low nutrient density. About 70% of the dry matter is starch. Potatoes are variable in composition, depending on variety, soil type, growing and storage conditions and processing treatment.

Expressed on a dry matter basis, whole potatoes contain about 60 to 120g per kg protein, 2 to 6g per kg fat, 20 to 50g per kg crude fibre and 40 to 70g per kg ash. Potato protein has a high biological value, among the highest of the plant proteins, and similar to that of soya beans.

The protein fraction in raw potato is poorly digested by poultry because of the presence of a powerful inhibitor of the gut enzymes that break down proteins. This inhibitor is destroyed by cooking. Consequently, potatoes should be cooked before being fed to poultry otherwise utilization is low and the droppings become very wet. Energy utilization as well as protein utilization is improved by cooking.

Another factor that needs to be considered is that potatoes may contain the glycoside solanin, which can result in poisoning particularly if the potatoes are green and sprouted. Consequently, green potatoes should be avoided for feeding.

There is a lack of recent information on the feeding of potatoes to poultry, but research with pigs has shown that cooked potato is approximately equal to maize in

energy and digestible protein contents. For best results potatoes should be boiled for 30–40 min, steamed at 100°C for 20–30 min or simmered for 1 hour, to ensure thorough cooking through to the centre, and then rapidly cooled. Prolonged heating, or slow cooling after heating, results in damage to the protein and a reduction in its digestibility.

Inclusion of cooked potatoes or potato flakes at levels up to 200g/kg in poultry grower diets has given good results in some investigations but growth appears to be depressed at higher levels. Similar findings have been reported with laying hens, egg production being significantly lower when the diet contained 200g/kg potato meal. Also, the droppings tended to be wetter. A dietary inclusion rate of 150g/kg potato meal has given satisfactory egg production. The wet droppings reported in these studies were probably due to the potassium content, which is known to be high in potatoes.

Molasses

Molasses is used mainly in feed mixtures for poultry as a pellet binder, therefore its nutritional composition is not of primary importance.

The amount used in organic poultry diets as a pellet-binding agent is 25–50g per kg. Molasses cause the feed granules to stick together during the pelleting process, resulting in pellets that are less likely to break down during transportation and passage through feeding equipment. Additional benefits of molasses addition are a possible increase in palatability of the diet and a reduction in dustiness of the dietary mixture. Higher dietary levels are known to have a laxative effect.

One problem with using molasses is that it is difficult to handle since it is a viscous syrup. Dry molasses meal is easier to handle and add to feed mixtures but may not be acceptable by all organic certifying agencies due to the chemicals added and the carrier used in converting the syrup to a dry form.

In tropical countries with a surplus of sugar cane the juice can be used as a feed source. Sugar cane juice has been fed to ducks with resulting rates of growth and feed conversion efficiency being only slightly inferior to those obtained with cereal-based diets.

FORAGES AND ROUGHAGES

Approved feedstuffs in this section include lucerne, lucerne meal, clover, clover meal, grass (obtained from forage plants), grass meal, hay, silage, straw from cereals and root vegetables for foraging. They can be fed fresh or conserved by haymaking, ensilage, etc.

Grass Meal

Poultry with access to outdoors will likely have access to grass. Also, in organic production grass (or other forage) meal should be included in diets for growing and laying poultry as a source of roughage and as a natural source of nutrients and carotenoids for skin and egg yolk pigmentation, during seasons when good forage is not available. Additionally, certain poultry, especially geese, are grazing birds in which intake of grass is a traditional part of the diet. Thus there is current interest in the utilization of grass and grass meal by poultry.

Grass meal is an established feed ingredient in Europe, based on research carried out more than 50 years ago. Bolton and Blair (1974) recommended that for use in poultry diets the product should contain at least 170g/kg protein.

Organic grass meal is likely to be a mixture of species including clover, sainfoin, etc., as well as true grasses and is therefore unlikely to be a consistent product. This factor has to be taken into account in assessing its suitability for feeding to poultry. Another main factor influencing the nutritive quality of grass meal is the stage of maturity at harvesting. The meal from grass cut at a young stage is a good source of protein, carotene and xanthophylls, riboflavin and minerals. Grass meal derived from the first, second and third cut of a pasture is known to have a reduced sugar content and an increased content of dietary fibre as the forage matures.

Research by Bolton and Blair (1974) indicated that that grass meal could be included in growing and laying poultry diets at levels up to 100g/kg. Higher levels have been used successfully in other studies. One interesting finding was that egg cholesterol content decreased as the level of lucerne or grass meal in the diet was increased, also that addition of grass meal to the diet increased egg hatchability in breeding birds.

Utilization of grass would be expected to be higher in geese because they are a grazing species, and this was confirmed by research showing that goslings grew as well on a diet containing 250g/kg chopped grass as on a grain-based, pelleted, diet.

Lucerne (Also known as Alfalfa)

This is a perennial herbaceous legume. Due to its high nutritional quality, high yields and high adaptability, lucerne is one of the most important legume forages of the world. It is used mainly for ruminant animals such as dairy cattle, beef cattle, horses, sheep and goats but is also used in poultry feed, both in processed form and as a forage.

Lucerne is usually cultivated for hay,

and is frequently used for silage or haylage, dehydrated to make meal or pellets, or used fresh by grazing or cut-and-feed. In several countries, dehydration plants produce lucerne pellets. Sun-cured lucerne has a protein content higher than in most cereal grains. It is high in calcium, low in phosphorus, and is a good source of other minerals and of most vitamins.

Dehydration is a good way to preserve lucerne for feed use, though the process is usually done off-farm. This preserves the maximum content of nutrients. The dehydrated product can subsequently be ground into a meal or made into pellets.

The protein content of dehydrated lucerne is about 170g/kg, making it a potential protein supplement for poultry diets. However, the fibre content is high at around 250 to 300g/kg, which makes the protein harder to digest. This limits the amount that can be included in poultry diets.

Seaweeds

Seaweeds (kelp) contain significant quantities of minerals but tend to be low in other nutrients. In some regions they are harvested as dried fodder for farm livestock in coastal areas. There is considerable potential for increased kelp production, particularly in regions such as the Pacific coast of North America.

The composition of seaweed differs according to species and to naturally occurring changes in the plant. Current research indicates that only low levels of seaweed can be included in organic poultry diets without affecting productivity.

Supplementation of the diet with seaweed has been shown to be of value in the control of parasites in pigs, markedly reducing the incidence of liver condemnations from ascarid damage in pigs at slaughter. This aspect of seaweed inclusion is of interest to

Component g/kg unless stated	Cassava	Molasses Beet	Molasses Cane	Potatoes cooked	Potato protein concentrate	Lucerne dehydrated	Grass meal
Dry matter	880	770	710	222	910	907	917
ME kcal/kg	3450	2100	2150	798	2840	660	1390
ME MJ/kg	14.4	9.5	9.6	3.36	11.7	2.1	5.8
Crude protein**	33	60	40	24.5	755	162	182
Lysine	0.2	0.1	0.1	1.43	64	8.1	7.1
Methionine + Cystine	0.9	0.1	0.5	0.7	32	5	5
Crude fibre	44	0	0	7.5	5.5	207	209
Crude fat	5	0	0	1.0	20	32	34.2
Linoleic acid	0.9	0	0	0.01	2.2	4.3	2.9
Calcium	2.2	2	8.2	0.14	2.5	12.7	6.3
Phosphorus	1.5	0.3	0.8	0.7	3.8	2.0	3.5
Available phosphorus	0.7	0.2	0.72	0.6	2.8	1.8	3.2

TABLE 4.3 Average composition of common tuber roots and forage by-products etc., for organic poultry feeding.*

* As-fed basis. ** N × 6.25

organic farmers but does not appear to have been investigated in poultry.

FEED MATERIALS OF MINERAL ORIGIN

The approved products to supply the mineral requirements of poultry include the following:

Sodium products: Unrefined sea salt, coarse rock salt, sodium sulphate, sodium carbonate, sodium bicarbonate and sodium chloride.

Calcium products: Lithotamnion and maerl shells of aquatic animals (including cuttlefish bones), calcium carbonate, calcium lactate, calcium gluconate.

Phosphorus products: Bone dicalcium phosphate precipitate, defluorinated dicalcium phosphate and defluorinated monocalcium phosphate.

The approved products to supply the trace element requirements include the following:

Iron products: Ferrous carbonate, ferrous sulphate monohydrate and ferric oxide.

Iodine products: Calcium iodate, anhydrous calcium iodate hexahydrate and potassium iodide.

Copper products: Copper oxide, basic copper carbonate monohydrate and copper sulphate pentahydrate.

Zinc products: Zinc carbonate, zinc oxide, zinc sulphate monohydrate and/or heptahydrate.

Selenium products: Sodium selenate and sodium selenite.

VITAMINS

The vitamins approved for use in organic feeding have preferably to be derived from ingredients occurring naturally in feeds, or synthetic vitamins identical to natural vitamins (related EU regulation, Directive 70/524/EEC). When the organic feed or organic poultry product is to be exported to the USA, the vitamins and trace minerals used have to be FDA-approved.

ENZYMES

Certain enzymes can be added to organic diets to help the release of nutrients during digestion (related EU regulation, Directive 70/524/EEC). Examples are:

1. Phytase, which acts on phytate phosphorus in plant materials, releasing more of the contained phosphorus. As a result less phosphorus supplementation is required and a lower amount of phosphorus is excreted in manure (possibly 30%).
2. β-Glucanase and xylanase, which can improve the digestibility of carbohydrate, fat and protein in some cereals.
3. α-Galactosidase, which may help the digestion of diets containing high levels of plant protein feedstuffs such as soya bean meal or lupin seed. These legumes contain oligosaccharides, which cannot be degraded by the gut enzymes and are fermented in the large intestine causing gas production.
4. Addition of α-amylase can improve the digestion of starch and addition of protease has been shown to improve the digestion of protein.

Producers considering the use of supplementary enzymes should consult a feed supplier and the local organic certifying agency before doing so.

MICROORGANISMS

Certain microorganisms are approved for use in organic poultry feed (related EU regulation, Directive 70/524/EEC).

Brewer's Yeast

This product (dried Saccharomyces cerevisiae) is permitted as a feed ingredient in organic diets, and has been used traditionally in animal diets as a source of nutrients. Inactivated yeast should be used for poultry feeding since live yeast may grow in the intestinal tract and compete for nutrients.

Other yeasts have been included in poultry diets and may be acceptable as ingredients for organic production.

OTHER ADDITIVES APPROVED FOR ORGANIC POULTRY FEEDING

The regulations permit the addition of certain additives for use in ensilage, as processing aids, pellet binders, etc. These are detailed in Chapter 2, Table 2.1.

Dried marigold petals and marigold extract do not appear to be listed in the above regulations but should be acceptable for use in poultry diets if grown and produced organically. This should be checked with the local organic certifying agency. Marigolds (*Tagetes erecta*) are rich in carotenoids (mainly lutein) and the petals and petal extracts are useful additives in poultry feed for colouring the skin, flesh (fat) and egg yolks. In the European Union, naturally derived lutein is classified as E161b, but marigold extract has not been assigned an E number and is traded as vegetable extract. In North America marigold meal and its extracts are approved as pigment sources in poultry feed, and marigold extracts have been approved for human food use (with FDA-GRAS status).

DIGESTION AND ABSORPTION OF NUTRIENTS

Poultry need a well-balanced and easily digested diet for optimal production of eggs and meat. They are very sensitive to dietary quality because they grow quickly and make relatively little use of fibrous, bulky feeds such as lucerne hay or pasture since, like pigs, they are non-ruminants and possess a simple stomach compartment.

Poultry have evolved as a seed-eating species, as can be concluded from the diagram of the digestive system (Fig. 4.1).

The digestive system can be seen as being relatively simple, related probably to an evolutionary need for a light body weight related to the ability to fly. The main features are:

1. A beak instead of a mouth with lips, and with no teeth to permit grinding of feed particles. The feed is not chewed but is swallowed whole. A sense of taste is not well developed in poultry.
2. A crop (pouch off the oesophagus). The crop serves as a storage organ to allow meal-eating. Unlike chickens, waterfowl do not possess a true 'crop' (an anatomical error made by Sir Arthur Conan Doyle in his story 'The Adventure of the Blue Carbuncle' in which a missing jewel is found hidden by Sherlock Holmes in the 'crop' of a goose). Instead, the oesophagus in waterfowl is capable of expanding to accommodate substantial amounts of feed.

In adult doves and pigeons, the crop produces 'crop milk', which is

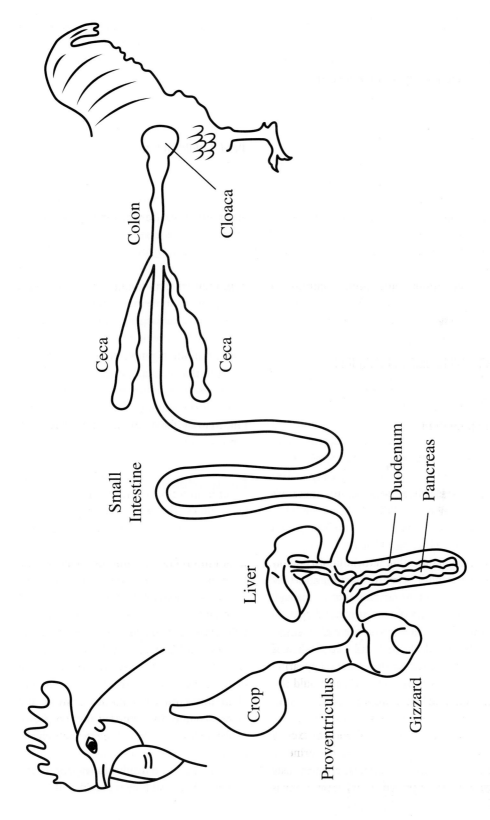

FIGURE 4.1 Digestive system of poultry.

regurgitated to feed newly hatched chicks. After a bird has eaten, the crop should appear full and bulge out from the chest wall when the bird is examined.

3. A proventriculus, where the feed particles mix with acid and digestive enzymes. This part of the gut functions like the glandular stomach in the pig.

4. A gizzard, which acts to grind up the ingested feed particles, assisted by the accumulation of insoluble grit and small stones.

5. A small intestine, where digestion of carbohydrates, proteins and fats is completed under the action of enzymes secreted by the pancreas and by bile produced in the liver. Absorption of nutrients into the bloodstream occurs in this organ.

6. A large intestine, where water is absorbed.

7. Two caeca at the terminal end of the large intestine, close to the cloaca (vent). These blind sacs contain microorganisms and function like the rumen in cattle and sheep, resulting in the breakdown of fibre that resisted digestion in the upper digestive tract. The caeca vary greatly in size in different types of birds, reaching their greatest size in those that are herbivorous and survive on diets containing high levels of fibre, e.g. ostriches.

8. In contrast to other farm animals, poultry do not have separate openings for the excretion of urine and faeces. Instead, all the excreta (faeces, urine and eggs, in the case of hens) pass through a single opening (cloaca or vent). Caecal waste is also deposited on the excreta, appearing as a light brown froth, which should not be confused as diarrhoea.

Chickens, like other birds, do not excrete liquid urine. Instead they excrete urine as uric acid, which is excreted as a white paste or a dry, white powder. Very little water is required for this process in birds, compared to the excretion of urine in cattle or pigs, and is related to their ancestry from reptiles. The process also explains the absence of a bladder in poultry.

A more detailed description of the digestive process in poultry is available in the book *Nutrition and Feeding of Organic Poultry* (Blair, 2008).

Owing to the high metabolic rate of the fowl, a more or less continuous supply of feed is required. This is provided for by the crop that acts as a reservoir for the storage of feed prior to its digestion.

Poultry tend to eat meals at about 15-minute intervals through the daylight hours and, to some extent, during darkness. They tend to eat larger portions at first light and in the late evening.

A meal of normal feed takes about 4 hours to pass through the gut in the case of young stock, 8 hours in the case of laying hens and 12 hours for broody hens. Intact, hard grains take longer to digest than cracked grain.

Feed Intake

Selection of feed is influenced by two types of factors: innate and learned. Although the chicken has relatively few taste buds and does not possess a highly developed sense of smell it is able to discriminate between certain feed sources on the basis of colour, taste and/or flavour. Learning to discriminate between nutritious and harmful feeds is helped in organic production by the presence of parent birds during the early life of the chick. Some research indicates that chicks have a preference for diets of the same colour as that provided after hatching.

According to recent evidence, chickens prefer yellow–white maize followed by yellow, orange and finally orange–red maize. Red, red–blue and blue seeds are eaten only when the birds are very hungry. Also, less was eaten of black and green diets. Rancidity and staleness are known to reduce intake of feed. It is not clear whether flavours added to poultry feed are effective in stimulating intake.

Sucrose in solution appears to be the only sugar for which chickens have a preference, indicating a way of alleviating or preventing 'starve-outs' in baby chicks or of helping birds during disease outbreaks or periods of stress.

Birds have been shown to possess some degree of 'nutritional wisdom' or 'specific appetites', eating less of diets that are inadequate in nutrient content. Laying stock (especially light breeds) have the ability to regulate feed intake according to the energy level of the diet. Therefore, it is important to adjust the concentration of other nutrients in relation to energy level so that intake of these nutrients is adequate.

Use can be made of this trait in planning choice feeding systems for poultry, as will be outlined in a later section.

Current research indicates that wheat and sunflower seeds, polished rice, cooked potatoes, potato flakes and fresh fish are very palatable feedstuffs for poultry. Oats, rye, rough rice, buckwheat and barley are less palatable, unless ground. Linseed meal has been found to be very unpalatable.

Particle size has an important effect on feed intake. One study showed that intake was greatest with particles between 1.18 and 2.36mm. As the birds aged the preference was for particles greater than 2.36mm.

Social interaction is another factor influencing intake, chicks being known to eat more in a group situation.

DIGESTIBILITY

As with other farm stock, only a fraction of each nutrient in the feed is absorbed from the digestive tract. Researchers can measure this by determining both the amount of nutrient present in the feed and the amount of nutrient present in the faeces, or more exactly in the ileum (the area in the small intestine where digestion is complete). The difference between the two, commonly expressed as a percentage (100 indicating complete digestion), is the proportion of the nutrient digested by the bird. Each feedstuff has its own unique set of digestibility coefficients for all nutrients present. The digestibility of an ingredient or a complete feed can also be measured.

Digestibility measured in this way is known as 'apparent digestibility', since the faeces and ileal digesta contain substances originating in the fluids and cellular material secreted by the gut and associated organs as the digesta pass. Correction for these endogenous losses allows for the 'true digestibility' to be measured. Generally, the digestibility values listed in feed tables refer to apparent digestibility unless stated otherwise.

Digestibility of Carbohydrates

Starch is the main energy source in poultry diets and is generally well digested. Complex carbohydrates such as cellulose, which represent much of the fibre in plants, cannot be digested by poultry. There is some microbial breakdown of cellulose in the caeca, at least in some avian species, which can increase the energy yield from the feed. The complex carbohydrates contained in barley, rye, oats and wheat increase the viscosity of the ingested feed, interfering with digestion and absorption. In addition, they also result in sticky droppings, which can lead to foot and leg problems and breast blisters. As a result, it is now a common practice to add the appropriate enzymes to conventional poultry diets to achieve breakdown of these components during digestion.

Chitin is the main component of the hard shell of insects and has a chemical structure like that of cellulose. Birds have some ability to digest this component but studies indicate that the insect skeleton is not an important source of nutrients for domestic poultry.

Some carbohydrate components in the feed may interfere with digestion. For instance, soya bean meal may contain a substantial level of oligosaccharides, which are indigestible in the upper intestinal tract of poultry (and pigs) due to the absence of the enzyme necessary for their breakdown, resulting in reduced digestibility of soya bean meal-based diets. Ways of addressing this issue include the use of low oligosaccharide cultivars of soya bean meal and addition of a specific enzyme to the feed.

Cooking improves the digestibility of some feedstuffs such as potato. Steam-pelleting may also improve starch digestibility.

Digestibility of Proteins

Proteins are broken down in the gut into amino acids, which are absorbed by the bird and used as building blocks to form feathers, eggs, muscle and sperm, etc. The more closely the released amino acids meet the amino acid needs of the bird the higher the quality of the protein supplied by the feed. Unused amino acids are utilised as energy sources and the nitrogen-containing components are excreted in the droppings.

Digestibility of Fats

Fats are insoluble in water and have to be emulsified in the gut by bile from the gall bladder before they can be broken down into fatty acids and other simple fat molecules by gut enzymes. These are then combined with protein and other compounds to allow their passage into the bloodstream for use throughout the body either as an energy source or as the building blocks of body fat, egg fats, etc. Poultry have a specific requirement for only one fatty acid – α-linoleic acid – which has to be present in the diet. Other fatty acids can be generated in the body of the bird.

Fat must be present in the diet for poultry to absorb the fat-soluble vitamins A, D, E, and K.

In addition to its role in nutrition, fat is commonly added to feed to reduce grain dust. Fat addition also improves the palatability of feed, making it more appetizing.

Fats, including those incorporated in feed, have a tendency to go bad, or become rancid. This is a year round problem, but the risk of feed going rancid is even greater in the summer. To prevent feed from going rancid, an approved antioxidant can be added to poultry diets containing added fat.

A main factor influencing fat digestibility is age, older birds being better able to digest fats than young birds. The effect of age is more apparent with saturated fats (such as tallow) than unsaturated fats (such as maize oil). Other factors that can influence fat digestibility include the amount of fat in the diet and presence of other dietary components, particularly minerals.

The addition of fat to the diet can reduce the rate of passage of feed through the gut. As a result, the ingested feed spends more time in contact with digestive enzymes, which enhances the extent of digestion of fat and non-fat components. This can result in the feed mixture having a higher energy value than can be accounted for from the sum of the energy value of the ingredients, and is known as the 'extra-caloric effect'.

A reduction in fat digestibility and of energy release of up to 30% due to oxidation of fat as a result of overheating during feed processing has been shown experimentally.

NUTRIENT REQUIREMENTS

An understanding of how well the products of digestion meet the needs for maintenance, growth and production is necessary so that suitable feed mixtures can be formulated.

Like all other animals, poultry require five components in their diet as a source of nutrients: energy, protein, minerals, vitamins and water. A nutrient shortage or imbalance in relation to other nutrients will adversely affect growth, health and production. Poultry need a well-balanced and easily digested diet for optimal reproduction and meat production. Like pigs, they are very sensitive to dietary quality because they grow quickly and make relatively little use of fibrous, bulky feeds such as lucerne hay or pasture.

Energy

Energy is produced when the feed is digested in the gut. The energy is then either released as heat or is trapped chemically and absorbed into the body for metabolic purposes. It can be derived from protein, fat or carbohydrate in the diet. In general, cereals and fats provide most of the energy in the diet. Energy in excess of requirement is converted to fat and stored in the body. The provision of energy accounts for the greatest percentage of feed costs.

The total energy (Gross Energy) of a feedstuff can be measured in a laboratory by burning it under controlled conditions and measuring the energy released in the form of heat. Digestion is never complete under practical situations, therefore measurement of Gross Energy does not provide accurate information on the amount of energy useful to the animal. A more precise measurement of energy is Digestible Energy (DE), which takes into account the energy lost during incomplete digestion and excreted in the faeces. The chemical components of feedstuffs have a large influence on DE values, with increased fat giving higher values and increased fibre and ash giving lower values. Fat provides about 2.25 times the energy provided by carbohydrates or protein.

More accurate measures of useful energy contained in feedstuffs are Metabolizable Energy (ME, which takes into account energy loss in the urine) and Net Energy (NE, which in addition takes into account the energy lost as heat produced during digestion). Balance experiments can be used to determine ME fairly readily from comparisons of energy in the feed and excreta, the excretion of faeces and urine together in the bird being a convenient feature in this regard. As a result, ME is the most common energy measure used in poultry nutrition in many countries. A more accurate assessment of metabolizable energy can be obtained by adjusting the ME value for the amount of energy lost or gained to the body in the form of protein nitrogen.

Several researchers have developed equations for the estimation of ME based on the chemical composition of the diet (Blair, 2008).

The requirements set out in this publication and taken mainly from the report on the Nutrient Requirements of Poultry (NAS–NRC, 1994) are based on ME, expressed as kilocalories (kcal) or megacalories (Mcal) per kg feed. This energy system is used widely in North America and in many other countries. Energy units used in some countries are based on joules (J), kilojoules (kJ) or megajoules (MJ). A conversion factor can be used to convert calories to joules, i.e. 1Mcal = 4.184MJ; 1MJ = 0.239Mcal; and 1MJ = 239kcal. Therefore, the tables of feedstuff composition in this publication show

ME values expressed as MJ or kJ as well as kcal per kg.

Protein and Amino Acids

The term protein usually refers to crude protein (measured as nitrogen content × 6.25) in requirement tables. Protein is required in the diet as a source of amino acids, which can be regarded as the building blocks for the formation of skin, muscle tissue, feathers, eggs, etc. Body proteins are in a dynamic state with synthesis and degradation occurring continuously, therefore a constant, adequate intake of dietary amino acids is required. An inadequate intake of dietary protein (amino acids) results in a reduction or cessation of growth or productivity and an interference with essential body functions.

There are 22 different amino acids (AAs) in the body of the bird, 10 of which are essential (EAA; arginine, methionine, histidine, phenylalanine, isoleucine, leucine, lysine, threonine, tryptophan, and valine), i.e. cannot be manufactured by the body and must be derived from the diet. Cystine and tyrosine are semi-essential in that they can be synthesized from methionine and phenylalanine, respectively. The others are non-essential (NEAA) and can be made by the body.

Methionine is important for feather formation and is generally the first limiting amino acid. Therefore, it has to be at the correct level in the diet. The level of the first limiting amino acid in the diet normally determines the use that can be made of the other EAA. If the limiting amino acid is present at only 50% of requirement then the efficiency of use of the other essential amino acids will be limited to 50%. This concept explains why a deficiency of individual amino acids is not accompanied by specific deficiency signs: a deficiency of any

EAA results in a generalized protein deficiency. The primary sign is usually a reduction in feed intake that is accompanied by increased feed wastage, impaired growth and production, and general unthriftiness. Excess amino acids are not stored in the body but are excreted in the urine as nitrogen compounds.

Although a protein requirement per se is no longer appropriate in requirement tables, stating a dietary requirement for both protein and EAA is a convenient way to ensure that all amino acids needed physiologically are provided correctly in the diet (NAS–NRC, 1994).

In most poultry diets, a portion of each amino acid that is present is not biologically available to the animal. This is because most proteins are not fully digested and the amino acids are not fully absorbed. The amino acids in some proteins such as egg or milk protein are almost fully bioavailable, whereas those in other proteins such as certain plant seeds are less bioavailable. It is therefore more accurate to express amino acid requirements in terms of bioavailable (or digestible) amino acids.

Protein and amino acid requirements vary according to the age and stage of development. Growing meat birds have high amino acid requirements to meet the needs for rapid growth and tissue deposition. Mature cockerels have lower amino acid requirements than laying hens, even though their body size is greater and feed consumption is similar. Body size, growth rate, and egg production of poultry are determined by the genetics of the bird in question. Amino acid requirements, therefore, also differ among types, breeds, and strains of poultry.

Dietary requirements for amino acids and protein are usually stated as proportions of the diet. However, the level of feed consumption has to be taken into account to ensure the total intake of protein and amino acids is appropriate. The protein and amino acid requirements derived by the NAS–NRC (1994) relate to poultry kept in moderate temperatures (18° to 24°C). Ambient temperatures outside of this range cause an inverse response in feed consumption; that is, the lower the temperature, the greater the feed intake and vice versa (NAS–NRC, 1994). Consequently the dietary levels of protein and amino acids to meet the requirements should be increased in warmer environments and decreased in cooler environments, in accordance with expected differences in feed intake. These adjustments are designed to help ensure the required daily intake of amino acids.

For optimal performance the diet must provide adequate amounts of essential amino acids, adequate energy and adequate amounts of other essential nutrients. The protein requirement values outlined by the NAS–NRC (1994) assume a maize–soya diet of high digestibility. It is advisable to adjust the dietary target values when diets based on feedstuffs of lower digestibility are formulated. The bioavailability of essential amino acids in a wide range of feedstuffs has been measured by researchers. The primary method has been to measure the proportion of a dietary amino acid that has disappeared from the gut when digesta reach the terminal ileum, using surgically altered birds. Interpretation of the data is, however, somewhat complicated. The values determined by this method are more correctly termed 'ileal digestibilities' rather than bioavailabilities because amino acids are sometimes absorbed in a form that cannot be fully used in metabolism. Furthermore, unless a correction is made for endogenous amino acid losses, the values are 'apparent' rather than 'true'.

The estimates of requirement are based on the assumption that the profile of dietary bioavailable essential amino acids should remain relatively constant during all growth stages, and that a slightly different profile is more appropriate for egg production. The desirable profile has been called Ideal Protein (IP). The protein need is minimized as the dietary essential amino acid pattern approaches that of ideal protein. The nearer the essential amino acid composition of the diet is to ideal protein, the more efficiently the diet is utilized and the lower the level of nitrogen excretion. Energy is also used most efficiently at this point, thus both protein and energy utilization are maximized.

Cereal grains, such as maize, barley, wheat and sorghum are the main ingredients of poultry diets and usually provide 30 to 60% of the total amino acid requirements. Other sources of protein such as soya bean meal and canola meal must be provided to ensure adequate amounts and a proper balance of essential amino acids.

Estimated amino acid requirements are shown in the tables at the end of this chapter, based on the concept of ideal protein (NAS–NRC, 1994).

Factors that affect the level of feed intake have an influence on requirements, a reduction in expected feed intake requiring the concentration of dietary amino acids to be increased. Correspondingly, the concentration of amino acids may be reduced when feed intake is increased.

Minerals

Minerals perform important functions in the body of the bird and are essential for proper growth and reproduction. In addition to being constituents of bone and eggs they take part in other essential processes. A lack of minerals in the diet can result in deficiency signs, including reduced or low feed intake, reduced rate of growth, leg problems, abnormal feather development, feather pecking and cannibalism, goitre, unthriftiness, breeding and reproductive problems, and increased mortality.

Poultry need at least 14 mineral elements and it is possible that other minerals may also be essential in the body. Under natural conditions it is likely that poultry can obtain part of their mineral requirements by ingesting pasture and pecking in the soil. However, these sources cannot be guaranteed to provide all the requirements consistently, therefore poultry diets must be supplemented with minerals.

Of the essential mineral elements, those likely to be deficient in poultry diets are calcium, phosphorus, sodium, copper, iodine, manganese, selenium and zinc.

Calcium and phosphorus are essential for the formation and maintenance of the skeleton. Together they make up more than 70% of the mineral content of the avian body, mainly combined with each other. These values indicate the importance of calcium and phosphorus in the diet. An inadequate supply of either one in the diet will limit the utilization of the other. Most of the calcium in the diet of the growing bird is used for bone formation, whereas in the mature laying bird most of the dietary calcium is used for eggshell formation.

A ratio of approximately 2 calcium to 1 non-phytate phosphorus (by weight) is appropriate for most poultry diets, with the exception of diets for laying hens and turkeys. A much higher level of calcium is needed for eggshell formation, and a ratio as high as 12 calcium to 1 non-phytate phosphorus (by weight) is more appropriate for layers. A deficiency of calcium is more likely than a deficiency of phosphorus. Cereal grains, which constitute most of the avian diet, are quite low in calcium although generally the calcium present in cereal grains and most feedstuffs is of higher availability than that of phosphorus. Legumes and pasture provide some calcium.

The phosphorus content of cereal grains and grain by-products is higher, although about one-half or more is in the form of organically bound phytate, which is digested poorly by poultry. Only about 10% of the phytate phosphorus in maize and wheat is digested by poultry. The phosphorus in animal products and phosphorus supplements is generally considered to be well utilized. The phosphorus in oilseed meals also has a low bioavailability. In contrast, the phosphorus in protein sources of animal origin is largely inorganic (meaning in this context not containing carbon; organic compounds are those containing carbon), and most animal protein sources (including milk and meat products) have a high phosphorus bioavailability. The phosphorus in dehydrated lucerne meal is highly available. Steam pelleting has been shown to improve the bioavailability of phytate phosphorus in some studies but not in others. The phosphorus in inorganic phosphorus supplements also varies in bioavailability. As a result, the requirements are now set out in terms of available phosphorus or non-phytate phosphorus.

An adequate amount of vitamin D is also necessary for proper metabolism of calcium and phosphorus.

Less is known about the availability of calcium in feedstuffs, but the level of calcium is generally so low that the bioavailability is of little consequence. The calcium in common supplementary sources such as ground limestone, oyster shell and dicalcium phosphate is highly available.

Six trace minerals have been shown to be needed as supplements in poultry diets; iron, copper, zinc, manganese, iodine and selenium. Feed suppliers are usually aware of deficient (and adequate) levels of the trace minerals present in local feedstuffs and will provide trace minerals mixes formulated appropriately.

Several studies have shown that omitting trace minerals from poultry diets depresses productivity and tissue mineral concentrations productivity during the early stage of growth and has progressively deleterious effects on productivity with increasing age of the birds. These findings are of importance to organic producers, in view of their relevance to production efficiency and product quality.

Vitamins

Although vitamins are required in small amounts, they serve essential functions in maintaining normal growth and reproduction. Few vitamins can be synthesized by the bird in sufficient amounts to meet its needs. Some are found in adequate amounts in the feedstuffs commonly used in poultry diets, others must be supplemented. Although the total amount of a vitamin may appear to be adequate, some vitamins are present in bound or unavailable forms in feedstuffs. Supplementation is then essential.

Poultry require 14 vitamins, but not all have to be provided in the diet.

The egg normally contains sufficient vitamins to supply the needs of the developing embryo. For this reason, eggs are one of the best animal sources of vitamins in the human diet.

Vitamin D can be synthesized in the body by the action of sunlight on a precursor (7-dehydrocholesterol) in the skin, which in summer can provide all the requirement for vitamin D in poultry housed outdoors. Research has shown that latitude and season both affect the amount of vitamin D produced; for instance no synthesis of vitamin D taking place during cloudless days in Boston (USA, 42.2 degrees N) from November through February. In Edmonton (Canada, 52 degrees N) this ineffective winter period extended from October through March. Further south (34 degrees N and 18 degrees N), sunlight was sufficient to allow vitamin D synthesis in the middle of winter. Presumably a similar situation prevails in the southern hemisphere.

These results demonstrate the dramatic influence of changes in solar UVB radiation on vitamin D3 synthesis in skin and indicate the effect of latitude on the length of the 'vitamin D winter' during which dietary supplementation of the vitamin is necessary for poultry housed outdoors. Organic poultry producers need to be aware of these findings. Without supplementation there is a seasonal fluctuation in body stores of the vitamin in poultry housed outdoors, requiring dietary supplementation during winter. Once this deficiency was recognized, dietary supplementation with vitamin D became common practice.

Response to Signs of Vitamin Deficiency

Vitamin deficiency signs are rarely specific. Thus, if a vitamin deficiency is suspected it is advisable to check with a nutritionist or veterinarian and add a supplement to the feed or, preferably, the drinking water (using a water-miscible preparation) since poultry do not eat well when deficient in some vitamins.

The prevailing organic standards may permit injection of vitamins to correct deficiencies, but this should be checked with the certifying agency.

Water

Water is also a required nutrient, the requirement being about 2–3 times the weight of feed eaten. The most important consideration with poultry is to ensure that there is an adequate supply of fresh, uncontaminated water available at all times. Water should always be available *ad libitum*, from drinkers designed for poultry. Water quality is important. The guidelines are based on Total Dissolved Solids (TDS) of up to 5,000mg per kg and pH between 6 and 8 being generally acceptable.

Birds are also very sensitive to the temperature of the drinking water, preferring cold water over water that is above the ambient temperature. This can affect feed intake.

PUBLICATIONS ON NUTRIENT REQUIREMENTS

There is currently no set of nutritional standards designed specifically for organic poultry. These standards will have to be derived from existing standards for commercial poultry.

Nutrient requirements in North America are based on the recommendations of the National Research Council–National Academy of Sciences, Washington, DC. The recommendations cover pigs, poultry, dairy cattle, horses, laboratory animals, etc., and are published as a series of books. The recommendations for each species are updated periodically, the current Nutrient Requirements of Poultry being the 1994, 9th revised edition. The information is used widely by the feed industry in North America and in many other regions. No comparable, up to date recommendations exist in other countries.

One of the limitations of published estimates of requirements is their general applicability. For instance, a main issue influencing nutrient requirements for energy and amino acids in the growing bird is the capacity for the genotype in question to deposit lean tissue as the bird grows to maturity or develops reproductive capacity. Responses to higher dietary concentrations of amino acids will be positive only in birds

with a genetic potential to deposit lean tissue rather than fat, or to produce a large number of eggs. As a result, it is difficult to establish nutritional standards for amino acids that can be applied generally to all genotypes. For this reason, the conventional broiler and layer industries in Europe, Asia, Australia and North America commonly use nutrient requirement models based on requirement data but tailored to specific strains and genotypes of poultry. These models require accurate information on input/output data and are beyond the scope of the average organic producer.

One of the criticisms of the NAS–NRC publications is that some of the data are old and out of date because the research in question was carried out some time previous to publication. Also, that the time lag in the derivation of new research findings, its peer review and publication in scientific journals, and its incorporation into the NAS–NRC recommendations makes the information less applicable to superior genotypes. This criticism is of less importance to organic producers who use traditional breeds and strains of poultry that have not been subjected to the selection pressure imposed on leading genotypes used in conventional production. Consequently, they should find the NAS–NRC publications a useful guide to nutrient requirements. It could be argued that, of the various requirement estimates available, e.g. the ARC (1975) estimates produced in the UK are the most applicable to organic production because of the genotypes used in deriving them. The data are, however, incomplete.

STANDARDS FOR DIETARY FORMULATION

Standards for organic poultry production can be derived from the NAS–NRC (1994) estimates of nutritional requirements, with the caveat that the requirement values estimated by NAS–NRC are designed for modern, highly productive birds with an increased potential for meat development and a high rate of egg production. The birds used in organic production are preferably heritage types, which develop more slowly and lay fewer eggs than modern hybrids. Also, the NAS–NRC values are based on high-energy maize–soya bean meal diets.

Therefore it is logical to modify the NAS–NRC values by lowering the dietary ME value to a level that is more suited to the energy needs of organic type of poultry, at least for chickens and turkeys. Reducing the energy content of the diet also minimizes the need for restricted feeding of breeder stock to avoid them becoming too heavy. In reducing the energy value of the diet it is logical to reduce the protein and amino acid contents by a similar proportion so that the same ratios of energy to protein and amino acids are maintained. This approach has been found to be successful in feeding organic turkeys.

However, it is recommended that the NAS–NRC (1994) derived values for mineral and vitamin requirements are adopted without modification, to help ensure correct skeletal growth and avoidance of foot and leg problems. Conventional diets are usually formulated with higher levels of minerals and vitamins but this approach is not suggested for organic diets, to try and minimize nutrient levels above those required for normal growth and reproduction.

Standards based on this approach, and derived feed formulas, are shown in the following tables. Producers using modern hybrids instead of traditional genotypes are advised instead to use the values suggested by the breeding company as the basis for diet formulation.

FEEDING ORGANIC POULTRY

A key aim of organic farming is environmental sustainability. Consequently, organic producers wish to provide most or all of their required inputs, including feed. This is not possible on small farms, and even larger farms that may produce some of the feedstuffs required may not have the necessary mixing equipment to allow adequate diets to be prepared on-site. Farms with a land base sufficient for the growing of a variety of crops may be able to mix diets on-site or in a co-operative feed mill.

Farms with an adequate supply of home-produced grains but no protein crops can purchase a supplement (sometimes called a concentrate) that provides all of the nutrients lacking in the grains. It should be purchased with an accompanying mixing guide. In North America the label information provided with the purchased supplement is similar to that provided above for complete feed. Various supplements are sold by feed manufacturers and farm stores, though perhaps only a single poultry supplement may be available and is expected to be used with all classes of birds. Use of supplement is therefore a compromise between convenience and accuracy in formulation. Any supplement purchased would have to comply with organic standards.

Below is an example formula for a supplement (320g crude protein/kg) that is designed for use with laying hens (Table 4.4).

One of the benefits of using a supplement is that it can be incorporated easily into a feeding programme based on the use of whole (unground) grain. This avoids the need for the farm to purchase mixing equipment.

Farms with adequate grain and protein

Ingredient	g/kg (90% DM basis)
Fishmeal	170
Soya bean meal	350
Grass meal	177
Ground limestone	230
Dicalcium phosphate	47
Salt (NaCl)	6.0
Trace mineral premix	10.0
Vitamin premix	10.0

TABLE 4.4 Example composition of a layer supplement (concentrate) (from Blair et al., 1973).

feedstuffs available for farm use can mix their own diets on-farm so that they have complete control over the formula. In this case it is necessary only to purchase a vitamin + trace mineral premix and also calcium, phosphorus and sodium (salt) sources.

Where possible, separate vitamin and trace mineral premixes should be used. Vitamins in contact with minerals over a prolonged period of time in a hot and humid environment lose potency, possibly resulting in vitamin deficiencies and reduced performance. In countries with high ambient temperatures it may be necessary to provide cooled storage facilities. When vitamins and minerals are combined in a single premix, it should be used within 30 days of purchase. Vitamin and trace mineral premixes should be stored in the dark in dry, sealed containers. Stabilizing agents are helpful in maintaining the quality of premixes, but must meet organic standards.

Separate premixes for laying, breeding and growing birds are recommended, since their requirements are different. The premixes supply all the required vitamins and trace minerals and assume no contribution from the main dietary ingredients

since these vary and may be low in potency or bioavailability. Producers who wish to include a lower level of premix on the basis that some vitamins and minerals are provided by the main ingredients should do so with care, observing the birds closely for signs of deficiencies and ensuring that their welfare is not compromised. Vitamin D can be omitted from summer formulations, provided the birds have access to adequate sunshine. Vitamin D supplementation is recommended in premixes for use in regions north of latitude 42.2 N and above, at all other times of the year. The carrier used in the premixes can be a material such as ground oat hulls.

Some producers prefer to use macro-premixes instead of premixes of the type suggested. These mineral premixes contain calcium, phosphorus and salt, in addition to trace minerals. Their inclusion rate in diets is determined largely by the varying need for supplementary calcium and phosphorus. Organic producers have much more control of the dietary formulation when the premix and major minerals are included in the diet separately. However, producers wishing

	Amount per kg premix	Supplies per kg diet
Trace minerals		
Copper	1g	5.0mg
Iron	16g	80.0mg
Manganese	12g	60.0mg
Selenium	30mg	0.15mg
Zinc	8g	40.0mg
Carrier	To 1kg	

TABLE 4.5 Composition of a trace mineral premix to provide the NAS–NRC (1994) estimated trace mineral requirements in chicken grower and layer diets when included at 5kg per tonne.

Note: Iodine is usually provided in the form of iodized salt, otherwise iodine should be included in the premix.

	Amount per kg premix	Supplies per kg diet
Vitamins, IU		
Vitamin A	500,000	2,500
Vitamin D$_3$	50,000	250
Vitamin E	2,000	10.0
Vitamins		
Biotin	30mg	0.15mg
Choline	260g	1300mg
Folacin	110mg	0.55mg
Niacin	6g	30mg
Pantothenic acid	2.0g	10mg
Riboflavin	750mg	3.75mg
Vitamins		
Cobalamin (Vitamin B12)	2mg	10.0µg
Carrier	To 1kg	

TABLE 4.6 Composition of a vitamin premix to provide the NAS–NRC (1994) estimated vitamin requirements in chicken grower and layer diets when included at 5kg per tonne.

Note: vitamin D3 can be omitted from the summer formula for birds with access to direct sunshine.

Stage	Starter 0–8 w	Grower 8–18 w	Laying/breeding hens
AME, kcal	2750	2600	2650
AME, MJ	11.5	10.9	11.1
Crude protein g	175	140	160
Lysine g	8.0	5.3	6.9
Methionine + cystine g	6.0	4.3	5.8
Calcium g	9.0	8.0	32.5
Phosphorus (non-phytate) g	4.0	3.5	3.5

TABLE 4.7 Recommended nutrient standards for organic layer chicken diets, amount per kg diet (from NAS–NRC, 1994).

the convenience of a combined mineral and trace mineral premix should adopt the option of using macro-premixes and follow the mixing directions provided by the supplier to prepare appropriate diets.

The section below explains the principles of feed formulation for those organic producers who chose to mix complete feeds on-farm. Nutrient composition in the example feed mixtures shown below cannot be guaranteed owing to ingredient variability. Producers using such formulas should have the diets analyzed periodically to ensure acceptability.

CHICKENS

Recommended nutrient standards for organic chicken diets that have been derived from the NAS–NRC (1994) values are shown in Table 4.7 and example organic feed

Ingredient, kg	Starter 0–8 w	Pullet grower 8–18 w	Laying/breeding hens
Barley	90	175	
Wheat	530	379	627
Wheatfeed	100	195	83
Oats		50	
Soya bean meal	150		175
Canola meal		50	
Field peas	100	100	25
Dehydrated lucerne or grass meal		25	
Ground limestone	14	13	77*
Dicalcium phosphate	12	9	9
Salt (NaCl)	3	3	3
Vitamin Premix	5.0	5.0	5.0
Trace mineral Premix	5.0	5.0	5.0
Total, kg	1000	1000	1000

TABLE 4.8 Examples of organic feed mixtures (air-dry basis) for layer chickens that meet the recommended standards.

* Supply as 1/3 granules, 2/3 powder.

mixtures that meet the recommended standards are shown in Tables 4.8–4.11. Some examples contain supplemental pure amino acids, which are allowed temporarily in some countries and make the feed mixtures that are cheaper, easier to formulate and superior from an environmental standpoint.

Feed mixtures prepared on-farm are usually in mash form, i.e. mixtures of coarsely ground cereal grains and protein sources, etc. These can be prepared using equipment such as concrete mixers or paddle mixers such as those used by bakers. Alternatively, specialised feed mixers can be used although they are more costly and may be difficult to justify on small farms.

Mash diets can be used at all stages for poultry but the particle size should be greater for older birds otherwise feed wastage may be excessive. Fine particles less than 600μm GMD (geometric mean diameter) should be avoided at all stages and generally it is

recommended that the optimum feed particle size is between 600 and 900μm.

Crumbled or pelleted feeds are preferred for older birds and pelleted feeds are recommended for use with any form of automated feeding system to prevent 'bridging' and blockages that occur with mash feeds.

Preparation of crumbled or pelleted feed is more costly than for mash feeds, therefore producers should take advantage of any requisite facilities offered by co-operatives.

Examples of other dietary formulations for growing and laying chickens that meet the standards for UKROFS (UK Register of Organic Food Standards, MAFF, London), EU and EU without supplemental amino acids are shown in the following tables (Tables 4.9 and 4.10, from Lampkin, 1997).

The formulations in Tables 4.9 and 4.10 illustrate an important point. The exclusion of pure amino acids from organic diets usually requires an excessive level of protein

Ingredients g/kg	Starter			Pullet grower	
	UKROFS	EU incl. amino acids	EU excl. amino acids	UKROFS	EU excl. amino acids
Grain	400	373	228	282	299
Wheat middlings	100	100	100	300	300
Brewers/distillers grains	–	4.0	–	126	5.0
Peas/Beans	150	150	106	18.0	101
Soya beans	178	167	317	–	–
Oilseeds	–	50	124	100	100
Dried grass/lucerne	50	50	50	100	100
Fishmeal	15	–	–	–	–
Vegetable oil	3.0	1.0	–	30.0	28.0
Yeast	39.0	36.0	18.0	3.0	15.0
Ca/P sources	30.0	33.0	27.0	16.0	19.0
Salt (NaCl)	29.0	30.0	28.0	23.0	31.0
Mineral/vitamin premix	3.0	3.0	3.0	2.0	2.0
Lysine/methionine	2.0	3.0	–	–	–
Calculated analysis g/kg unless stated					
CP	211	201	250	176	150
ME MJ/kg	11.5	11.5	11.5	11.0	11.0
Lysine	13	13	16	7.0	8.0
Methionine	6.0	6.0	5.0	3.0	3.0
Linoleic acid	17.0	18.0	22.0	29.0	29.0
Calcium	12.0	12.0	12.0	8.0	8.0
Non-phytate P	5.0	5.0	5.0	5.0	5.0

TABLE 4.9 Example diets for organic pullet feeding in the UK (from Lampkin, 1997).

to be included in the diet (in the starter diet example, 250g/kg instead of 201 or 211 when pure amino acids can be used). This has a marked effect on increasing the cost of the diet. In addition, it results in an inefficient use of scarce protein supplies as well as contributing to an increased output of nitrogen in manure, a potential environmental hazard.

Bennett (2006a) published examples of organic diets designed for small-scale egg producers, with and without supplementation with pure amino acids (Table 4.11).

The diet with 160g protein/kg was recommended for feeding from the onset of lay until egg production dropped to 85%, after which the diet with 140g protein/kg was recommended.

All these formulas are very useful in allowing producers to select the appropriate type of feed mixture to adopt in order to comply with local organic standards, especially those with home-grown supplies of grain and protein feedstuffs. Bennett (2006a) noted that diets without supplementation with pure methionine were likely

Ingredients g/kg	UKROFS	EU incl. amino acids	EU excl. amino acids
Grain	202	303	237
Wheat middlings	300	297	300
Brewers/distillers grains	63	6	–
Peas/Beans	148	150	150
Soya beans	–	–	63
Dried grass/lucerne	50	50	50
Vegetable oil	77	34	36
Yeast	36	50	45
Ca/P sources	92	82	87
Salt (NaCl)	29	25	29
Mineral/vitamin premix	3	2	2
Lysine/methionine	1	1	–
Calculated analysis g/kg unless stated			
CP	160	160	170
ME MJ/kg	11	11	11
Lysine	8	8	10
Methionine	3	3	3
Linoleic acid	49	27	31
Calcium	35	35	35
Non-phytate P	5	5	5

TABLE 4.10 Example diets for organic layer feeding in the UK (from Lampkin, 1997).

	Diet with 160g/ kg CP	Diet with 140g/ kg CP	Soya bean meal available	Soya beans and peas available
Ingredients g/kg				
Wheat	474	561	744	526
Peas	333	327	–	220
Soya beans, cooked	77	–	–	147
Soya bean meal	–	-	150	–
Ground limestone	92	92	84	83
Dicalcium phosphate	14.3	10.8	10.5	11.2
Salt (NaCl)	3.1	2.7	2.6	3.0
DL-methionine	1.6	1.5	–	–
Vitamin/mineral premix	5.0	5.0	10.0	10.0

TABLE 4.11 Example dietary mixtures for organic layer feeding in Canada (from Bennett, 2006a).

to result in smaller and fewer eggs and an increased incidence of stress and cannibalism in the layers.

A general recommendation is that the birds should be allowed free access to the diets at all stages. The feed for growing pullets and laying hens not on pasture should be supplemented with good forage, to ensure an acceptable skin and yolk colour.

Meat Chickens

The NAS–NRC (1994) estimated the nutrient requirements of modern, hybrid-type broiler chickens (Table 4.12, abbreviated), which may be of interest to some organic producers. However, it is more likely that dietary specifications tailored to slower growing, more traditional genotypes are preferred by organic producers.

Recommended standards for use in organic production are shown in Table 4.14. These standards are based on Label Rouge standards for pastured birds and are designed to produce slower growth in the birds so that they are suitable for marketing at 12 weeks rather than 6–8 weeks, and have fewer health problems such as ascites, leg problems and sudden death syndrome which are common in commercial production.

Lewis et al. (1997) compared the

	Starting 0–3 w	Growing 3–6 w	Finishing
ME kcal a	3200	3200	3200
ME MJ	13.4	13.4	13.4
Crude protein g	230	200	180
Lysine	11.0	10.0	8.5
Methionine + cystine	9.0	7.2	6.0
Minerals g			
Calcium	10.0	9.0	8.0
Phosphorus (non-phytate)	4.5	3.5	3.0

TABLE 4.12 NAS–NRC (1994) estimated nutrient requirements of broiler chickens, amounts per kg diet (900g/kg DM basis)[a].

[a] Typical ME level used in conventional diets. Some values were stated as being tentative.

Stage	Starter 0–4 w	Grower 4–8 w	Finisher 8–12 w
AME, kcal	2900	2900	2930
AME, MJ	12.12	12.16	12.27
Crude protein g	200	180	160
Lysine g	10.6	8.8	7.2
Methionine + cystine g	6.8	6.0	5.5
Calcium g	9.0	8.9	8.8
Phosphorus (non-phytate) g	4.5	4.4	4.4

TABLE 4.13 Recommended nutrient standards for organic pastured meat-type chicken diets, based on Label Rouge standards, amount per kg diet.

Ingredient, kg	Starter 0–4 w	Grower 4–8 w	Finisher 8–12 w
Maize	50	50	65
Wheat	600	650	650
Wheatfeed	-	25	85
Soya bean meal	310	235	160
Ground limestone	12	12	12
Dicalcium phosphate	15	15	15
Salt (NaCl)	3	3	3
Vitamin premix	5.0	5.0	5.0
Trace mineral Premix	5.0	5.0	5.0
Total, kg	1000	1000	1000

TABLE 4.14 Examples of organic feed mixtures (air-dry basis) for pastured meat-type chickens that meet) the recommended standards.

Ingredient g/kg	Starter	Grower	Finisher
Maize	400	400	400
Wheat	220	250	300
Wheat middlings	70	110	120
Soya bean Hipro meal	280	210	150
Limestone flour	3.0	3.0	3.0
Dicalcium phosphate	14.0	14.0	14.0
Sodium bicarbonate	2.2	2.2	2.2
Salt (sodium chloride)	2.2	2.1	2.1
DL-methionine	1.0	1.0	0
Vitamin/drug premix	10.5	10.5	10.5
Calculated analysis g/kg (unless stated)			
ME MJ/kg	12.12	12.16	12.27
ME kcal/kg	2900	2900	2930
Crude Protein	202	178	156
Lysine	10.6	8.8	7.2
Methionine	3.3	2.9	2.6
Crude fibre	31.0	32.0	32.0
Calcium	9.0	8.9	8.8
Non-phytate phosphorus	4.5	4.4	4.4

TABLE 4.15 Composition and calculated nutrient content of broiler diets formulated to Label Rouge standards (Lewis et al., 1997).

production of broilers fed to Label Rouge standards. The compositions of the diets are shown in Table 4.15.

The diets did not claim to be organic but could be formulated to organic standards quite readily. Surprisingly the diets contained the anticoccidial drug salinomycin, also the amino acid pure methionine, indicating that these additives are permitted in the Label Rouge system.

Besides broilers, Label Rouge standards exist for layers, turkeys, ducks, geese, guinea fowl and capons. Layers require double yards (rested in rotation) because they are on range longer than broilers.

Lampkin (1997) outlined a typical feeding programme in the UK for meat birds, based on diets with a reduced content of energy, protein and amino acids to result in slower growth than in conventional production. Examples of formulations are shown in table 4.16.

	Starter		Grower		Finisher	
Ingredients g/kg	UKROFS	EU excl. amino acids	UKROFS	EU excl. amino acids	UKROFS	EU excl. amino acids
Grain	450	312	250	143	550	614
Wheat middlings	100	100	300	300	-	-
Maize gluten meal	-	-	-	-	-	85.0
Brewers/distillers grains	24.0	-	-	-	5.0	-
Peas/Beans	100	100	100	37.0	100	-
Soya beans	107	238	137	270	153	175
Oilseeds	-	108	7.0	98.0	14.0	91.0
Dried grass/lucerne	50.0	50.0	50.0	50.0	50.0	13.0
Fishmeal	64.0	-	16.0	-	-	-
Vegetable oil	32.0	-	50.0	28.0	29.0	3.0
Yeast	35.0	33.0	33.0	19.0	37.0	50.0
Ca/P sources	13.0	25.0	23.0	23.0	29.0	29.0
Salt (NaCl)	22.0	31.0	30.0	30.0	27.0	27.0
Mineral/vitamin premix	3.0	3.0	2.0	2.0	3.0	4.0
Lysine/methionine	1.0	-	2.0	-	1.0	-
Calculated analysis g/kg unless stated						
CP	207	238	189	220	171	205
ME MJ/kg	12.0	12.0	12.0	12.0	12.0	12.0
Lysine	13.0	14.0	11.0	14.0	11.0	10.0
Methionine	5.0	4.0	5.0	4.0	4.0	3.4
Linoleic acid	29.0	19.0	41.0	37.0	29.0	18.0
Calcium	10.0	10.0	10.0	10.0	10.0	10.0
Non-phytate P	5.0	5.0	5.0	5.0	5.0	5.0

TABLE 4.16 Example diets and nutrient specifications for free-range broiler feeding in the UK (from Lampkin, 1997).

	Starter 180g/kg CP	Finisher 140g/kg CP Peas and soya beans available	Finisher 140g/kg CP Peas available	Single diet A No supplemental amino acids	Single diet B No supplemental amino acids
Ingredients g/kg					
Wheat	561	760	667	768	578
Peas	250	100	293	–	158
Soya beans, cooked	146	100	–	–	222
Soya bean meal	–	–	–	192	–
Ground limestone	14.1	14.4	14.5	10.8	11.5
Dicalcium phosphate	18.6	15.9	16.0	17.1	17.1
Salt (NaCl)	3.0	2.9	3.1	2.0	2.4
L-lysine HCl	0.5	0.9	0.4	–	–
DL-methionine	1.9	0.5	1.0	–	–
Vitamin/mineral premix*	5.0	5.0	5.0	10.0	10.0
Enzyme mixture	0.5	0.5	0.5	+	+

TABLE 4.17 Example diets for organic roaster (broiler) feeding in Canada (from Bennett, 2006a).

* Level recommended by supplier.

Bennett (2006a) published examples of organic diets designed for small-scale producers of meat-type chickens, with and without supplementation with pure amino acids (Table 4.17). The diets were formulated to provide a lower level of protein than in conventional production. This researcher observed that diets without supplementation with pure methionine contained an imbalance of amino acids, resulting in slower growth and poor feathering until the birds were 6–8 weeks of age. The programme involved roaster starter to be used for the first 4 weeks, followed by a 50:50 mixture of starter and finisher for the next 2 weeks. From then until market the birds were given finisher diet. Example formulas for a simplified single diet to be fed from start to marketing (without supplemental amino acids) were also provided.

TURKEYS

Recommended nutrient standards for organic turkey diets are shown in Table 4.18, based on the NAS–NRC (1994) values. Example organic feed mixtures that meet (or exceed) the recommended standards are shown in Table 4.19. As with the chicken formulations, some contain supplemental pure amino acids.

'Starve-outs' are common in starting turkey poults, even though the feed is usually available on material such as egg flats. Sprinkling the feed with a green feedstuff such as grass or lucerne will help to attract the poults to feed. Tapping the feeder also helps. The presence of older poults that have already learned to feed is a more certain way to teach the young poults to feed successfully, or by placing some chicks with

Stage	Starter 0–6 weeks	Grower 6–10 weeks	Finisher 10 weeks–market	Breeding hens
AME, kcal	2600	2700	2900	2900
AME, MJ	10.9	11.3	12.1	12.1
Crude protein g	260	198	150	140
Lysine g	14.9	11.7	7.3	6.0
Methionine + cystine g	10	7.2	5	4.0
Calcium g	11.0	8.5	7.5	22.5
Phosphorus (non-phytate) g	5.5	4.5	4	3.5

TABLE 4.18 Recommended nutrient standards for organic turkey diets, amount per kg diet.

Ingredient, kg	Starter		Grower	Finisher	Layer/ Breeder
	AA	No AA	No AA	No AA	No AA
Maize	80	100	140		150
Wheat	276	233	285	628	500
Wheatfeed	35	30	141	121	100
Soya bean meal	450	481	340	140	
Soya beans cooked					100
Peas, field	75	55	65	25	55
Faba beans			–	60	30
Fishmeal, white	50	90	–		–
DL-methionine	2				
Ground limestone	18	6	11	10	51
Dicalcium phosphate	12	4	14	12	10
Salt (NaCl)	1	0	3	3	3
Vitamin premix	5.0	5.0	5.0	5.0	5.0
Trace mineral Premix	5.0	5.0	5.0	5.0	5.0
Total, kg	1000	1000	1000	1000	1000

TABLE 4.19 Example organic feed mixtures (air-dry basis) for turkeys that meet the recommended standards, with (AA) and without (no AA) supplemental methionine.

the poults. The chicks can be removed once the poults are eating and drinking well.

Another approach is to hatch and brood the poults naturally, though this is usually only feasible with very small flocks.

Grower and finisher diets are introduced as the poults grow to market weight. A feature of these diets is a gradual increase in energy value, aimed at producing an acceptable degree of finish (subcutaneous fat) on the birds at market weight.

DUCKS

Recommended nutrient standards for growing and breeding ducks are shown in Table 4.20. Nutrient requirements are not

Stage	0–2 weeks	2 weeks to market	Breeding
AME, kcal	2750	2650	2650
AME, MJ	11.5	11.1	11.1
Crude protein g	190	140	160
Lysine g	9.0	7.0	7.0
Methionine + cystine g	7.0	5.5	6.0
Calcium g	6.5	6.0	27.5
Phosphorus (non-phytate) g	4.0	3.0	3.5

TABLE 4.20 Recommended nutrient standards for growing and breeding ducks (White Pekin) diets, amount per kg diet (based on NAS–NRC, 1994, air-dry basis).

well defined. Diets can be formulated to meet these standards, but many producers find it easier to use chicken diets.

Ducks are grazers, like geese. The use of pelleted diets avoids spoilage and wastage. If duck feeds are not available, ducklings can be fed on chick starter for the first 2 to 3 weeks. The feed should be placed for the first few days on egg case flats or other rough paper to avoid leg injuries on slick-surfaced paper. After 2 to 3 weeks the ducklings can be fed a pelleted chicken grower diet plus cracked maize, or other grain. It is preferable to provide the feed in pellet form since mash has a tendency to stick in the bills of ducks and may cause them to choke, especially when very young. Also, crumbled or pelleted feeds will result in less feed wastage and better feed efficiency. Grower-size insoluble grit should be provided.

Lighting in the duck shelter is recommended so that ducklings can eat and drink at night. Lighting also helps to prevent the ducklings from becoming frightened, since the young birds have a highly nervous nature unless the mother duck is present.

Small flocks of ducklings raised in the late spring with access to green feed outdoors generally have few nutritional problems. While ducks are not as good foragers as geese, they do eat some green feed when allowed access to forage. Water for swimming is not essential for successful duck production but is often provided to allow proper bill cleaning.

In general, ducks are smaller and require less space than geese. The domesticated duck requires a grain supplement all year round, while geese can do well with limited grain supplementation provided they have access to good pasture.

Under organic conditions, Pekin ducklings are ready for market when about 8–10 weeks old.

GEESE

The nutritional requirements for geese are not well defined, especially since geese are foragers and are better able to digest fibre than chickens. Recommended nutrient standards are shown in Table 4.21, similar to those for chickens.

Commonly, the goslings are given starter feed for about 3 weeks when fed rationed amounts so that by 2 months of age good grass should be providing much of their diet. By 4 months, depending on the quantity and quality, geese can survive on grass alone with grain supplementation when needed. By 4 months of age they

Stage	0–3 weeks	3–18 weeks	Breeding
AME, kcal	2750	2600	2900
AME, MJ	11.5	10.9	12.1
Crude protein g	175	140	150
Lysine g	8.0	5.3	6.9
Methionine + cystine g	6.0	4.3	5.8
Calcium g	9	8	22.5
Phosphorus (non-phytate) g	4	3.5	3.5

TABLE 4.21 Recommended nutrient standards for growing and breeding goose diets, amount per kg diet.

can grow well on good pasture alone, with grain supplementation when needed, and be suitable for marketing at about 18 weeks of age. When pastures are of low quality the grower diet should be supplemented with dried forage.

A chicken breeder diet can be fed during the breeding period, or with a slightly lower level of calcium when the geese have access to a feeder containing limestone granules or oyster shell grit.

Research indicates that geese prefer clovers, bluegrass, orchard grass, timothy, and brome grass over lucerne and narrow-leaved tough grasses. This trait makes them suitable for biological control of weeds in crops such as lucerne.

Geese have a loud, harsh call when startled so they are sometimes used as 'watch' or guard geese. Geese can be quite aggressive and less likely to be attacked by predators than other types of poultry.

There is a growing niche market for goose eggs, which have a high percentage of egg white and are high in protein but have a higher content of cholesterol than chicken or duck eggs. Down is also an important product of goose farming.

PHEASANTS

Recommended nutrient standards for growing and breeding pheasants (and quail) are shown in table 4.22. Producers with flocks that are sufficiently large may opt to

Stage	0–4 weeks	5–8 weeks	9–17 weeks	Breeding
AME, kcal	2800	2800	2700	2800
AME, MJ	11.7	11.7	11.3	11.7
Crude protein g	280	240	180	180
Lysine g	17.0	14.0	8.0	10
Methionine + cystine g	10.0	9.3	6.0	6.5
Calcium g	10.0	8.5	7.5	25.0
Phosphorus (non-phytate) g	5.5	5.0	4.5	4.0

TABLE 4.22 Recommended nutrient standards for growing and breeding pheasants (ring-necked) diets, amount per kg diet.

formulate diets to these specifications, but generally it is more convenient to use turkey diets.

During the first 6 to 8 weeks of age, quail can be fed a starter feed containing 280g/kg protein, in crumble or mash form. From 8 to 14 weeks, a grower feed containing 200g/kg protein can be used, pelleted or crumbled. Some of the grain portion should be provided in whole form, also in the breeder diet.

RATITES (OSTRICHES, EMUS AND RHEAS)

Although native to Africa (ostriches), Australia (emus) or South America (rheas) these species are adaptable in environmental requirements and are now being farmed in several countries.

The nutrient requirements of these species have not been reviewed by the NAS–NRC (1994) or by any other similar group. Recommendations on feeding standards have, therefore, to be derived partly on research data on feedstuffs utilisation in these species and on practical experience (Table 4.23). Another factor making precise standards difficult is that ostriches are known to obtain more energy from feedstuffs than chickens, especially those feedstuffs with higher contents of fibre. This is related to the digestive system of ostriches, which is better equipped to digest fibre and helps to stimulate their natural grazing behaviour.

Producers with sufficiently large flocks may wish to use a series of diets, tailored specifically to stages in the life cycle of ratites and based on the recommended data in Table 4.23. Examples of feed mixtures that meet the recommended standards for ostrich feeding are shown in Table 4.26.

An alternative feeding programme that

has given good results in the USA involves feeding the chicks 24 to 48 hours after hatching on a good quality turkey or game bird starter feed containing at least 180g/kg protein for the first 2 weeks. Refusal to eat and drink is a common problem with young ostrich chicks while they are absorbing the yolk sac. This problem can be alleviated by placing several older chicks (1 to 3 weeks of age) that are already eating with the younger chicks. If older chicks are not available, domestic poultry chicks can be placed with newly hatched ostrich chicks to teach them to eat and drink. Access to continuous light as well as to the starter feed at all times during the first 3 weeks is recommended. After that time the birds can be fed all they will consume in two short (20 minute) daily feeding periods. Good quality small alfalfa pellets should be available to chicks on a continuous basis when twice daily feeding of the starter diet commences. This stimulates development of the gizzard and minimizes excessive weight gain, which can contribute to leg weakness problems. At 6 weeks of age the chicks can be introduced gradually to good quality ratite, game bird, or turkey grower diet and be allowed to consume all they will eat in two daily feeding periods. Lucerne pellets should continue to be offered as a supplemental feed unless good quality forage is available. The birds can be maintained on this feeding programme until they reach sexual maturity. A breeder feed should be fed to the laying/breeding flock, along with a supplemental source of calcium.

A single diet for all ages of ostriches is often more practical than separate starter, grower and breeder feeds, especially since the numbers farmed in any age class are generally too low to justify several diets.

Suggested specifications for such a diet are shown in Table 4.24 (Ullrey and Allen, 1996). The diet can be pelleted and fed *ad*

Stage	Pre-starter 0–2 weeks	Starter 2–6 weeks	Grower/Finisher 6 weeks–maturity/market	Laying/Breeding hens
AME, kcal	As for turkey starter	2500	2500	2300
AME, MJ		10.5	10.5	9.6
Crude protein g		180	170	200
Lysine g		7.5	7	9
Methionine + cystine g		5	4.5	7
Calcium g		14	13.5	24
Phosphorus (non-phytate) g		7.2	6.5	7
Crude fibre (max.)		65	110	100

TABLE 4.23 Recommended nutrient standards for organic ratite diets, amount per kg diet.

Nutrient specifications (90% DM basis)	g/kg unless stated
ME	2500kcal/kg
ME	10.5MJ/kg
Crude Protein	220
Lysine	12
Methionine + cystine	7.0
Crude fibre	100
Calcium	16.0 b
Non phytate Phosphorus	8.0

TABLE 4.24 Suggested nutrient specifications for a single, life cycle diet[a] suitable for ostriches, rheas, and emus (based on Ullrey and Allen, 1996).

[a] To be provided in pelleted form, *ad libitum*. b Additional calcareous grit or oyster shell grit to be provided for breeders.

libitum. According to the authors, the concentrations of energy, protein and essential amino acids in this diet may limit body weight gain somewhat below maximum, but are sufficient to support proportional development of the skeleton without excessive soft tissue weight in growing birds. It is therefore appropriate for organic production. Use of a feeder containing calcareous grit or oyster shell can be used to ensure a sufficient intake of calcium by laying/breeding females.

Scheideler and Sell (1997) published guidelines on feeding ostriches (Table 4.25) and emus (Table 4.27), which can be used as standards.

One feature of these recommendations is that prescribed levels of fibre are suggested, in keeping with the natural ability of these birds to digest fibrous feedstuffs more efficiently than chickens or turkeys. The diet of ostriches in the wild consists mainly of green grasses, berries, seeds, succulent plants and small insects. Using good quality grass or lucerne meal as the source of fibre is important, at least in the starter diet, since the green colour stimulates pecking and feed intake, and helps to reduce feather pecking.

g/kg diet unless stated (air-dry basis)	Starter 0–9 weeks	Grower 9–42 weeks	Finisher 42 weeks–market	Holding 42 weeks to sexual maturity	Laying from 4–5 weeks prior
AME kcal	2465	2450	2300	1980–2090	2300
AME MJ	11.1	11.1	9.6	8.3–8.7	9.6
Crude protein	220	190	160	160	200–210
Lysine	9.0	8.5	7.5	7.5	10.0
Methionine + cystine	7.0	6.8	6.0	6.0	7.0
Crude fibre	60–80	90–110	120–140	150–170	120-140
Calcium	15.0	12.0	12.0	12.0	24.0–35.0
Phosphorus (non-phytate)	7.5	6.0	6.0	6.0	7.0

TABLE 4.25 Recommended nutritional guidelines for ostrich diets (from Scheideler and Sell, 1997).

Ingredient, kg	Starter 2–6 weeks	Finisher 9 weeks–maturity/market	Laying/Breeding hens
Barley			66
Maize	250	190	–
Sorghum	200		265
Wheat	–	295	
Wheatfeed	110	167	123
Oats	75		
Soya bean meal	150	100	360
Canola meal	–	99	–
Groundnut meal	100		25
Dehydrated lucerne or grass meal	50	90	75
Ground limestone	9	11	43
Dicalcium phosphate	43	35	30
Salt (NaCl)	3	3	3
Vitamin premix	5.0	5.0	5.0
Trace mineral Premix	5.0	5.0	5.0
Total, kg	1000	1000	1000

TABLE 4.26 Example organic feed mixtures (air-dry basis) for ostriches that meet the recommended standards.

g/kg diet unless stated (air-dry basis)	Starter 0–6 weeks	Grower 6–36 weeks	Finisher 36–48 weeks	Breeder holding 42 weeks–sexual maturity	Breeder from 4–5 weeks prior
AME kcal	2685	2640	2860	2530	2400
AME MJ	11.23	11.05	11.97	10.59	10.04
Crude protein	220	200	170	160	200–220
Lysine	11.0	9.4	7.8	7.5	10.0
Methionine + cystine	4.8	4.4	3.8	3.6	4.0
Crude fibre	60–80	60–80	60–70	60–70	70–80
Calcium	15.0	13.0	12.0	12.0	24.0–35.0
Phosphorus (non-phytate)	7.5	6.5	6.0	6.0	6.0

TABLE 4.27 Recommended nutritional guidelines for emu diets (from Scheideler and Sell, 1997).

Since emus are grown mainly for their oil, it is probably logical to feed finisher diets with a higher content of energy and a lower content of fibre than for ostriches, to stimulate deposition of body fat. Also, in contrast to ostrich breeders, emu breeders need to have ample energy reserves (fat pad) in the body at the onset of breeding. Feed intake by emus decreases and becomes erratic during reproduction, and body fat is needed to carry them through in reasonable condition. Hence the differences in the recommendations shown in Tables 4.25 and 4.27 (Scheideler and Sell, 1997).

Detailed information on the nutritional requirements of rheas is very limited, and sound feeding management strategies have not yet been established for pastured birds.

CHOICE FEEDING USING WHOLE CEREAL GRAINS

Feeding any of the feed mixtures outlined above will ensure that the birds in question are receiving an appropriate combination of nutrients.

However, an alternative system for organic producers to adopt is choice feeding, involving the use of whole grain that may be available on-farm. This system approaches the natural feeding system more closely than other feeding systems and is therefore highly appropriate for organic production.

Also, one of the disadvantages of feeding a complete diet is that birds can only adjust intake according to their appetite for energy. When the environmental temperature falls or rises, the birds either over or under consume protein and minerals such as calcium.

The feeding of whole grain to poultry was used commonly in the past in many countries before commercial poultry production became intensified, the chickens being allowed to forage and be fed mainly on scratch grains. However, this earlier feeding system was largely abandoned in favour of all-mash (ground) or pelleted diets when large-scale, intensive production using automated feeding systems and high-producing stock was introduced.

As pointed out by Blair (2008), the bird has a digestive system designed to digest

FIGURE 4.2 Choice feeding of poultry.

whole seeds, therefore it seems illogical and unnecessary to feed it a pre-ground diet. Feeding whole grain also presents energy savings in feed preparation. The process of grinding requires about 20kWh/tonne of grain, while pelleting requires a large input of electrical energy amounting to about approximately 10% of the total feed costs and additional energy is required to generate steam for the steam-pelleting process.

Another advantage of whole grain feeding is that it has been shown to result in health benefits by reducing leg and skeletal problems, through a reduction in early weight gain and by promoting a better developed gizzard, which appears to act as a barrier organ in preventing pathogenic protozoa (causing coccidiosis) and bacteria from entering the lower digestive tract.

A main aim in feeding organic poultry should, therefore, be to maximise the use of whole grains. This can be achieved practically by allowing the birds access to whole grain and a supplement in separate feeders. The supplement is designed to supply all the nutrients lacking in grain.

Under choice feeding, or 'free choice feeding', birds are usually offered a choice of up to three types of feed: (a) an energy source (grain); (b) a protein source (protein mixture supplemented with vitamins and minerals) and (c), in the case of laying hens, calcium in granular form (calcareous grit such as oyster shell grit). Regular grit of a suitable granular size (to aid grinding in the gizzard) should also be available.

The basic principle behind choice feeding is that birds possess some degree

Feed ingredient offered	Intake % of total	Nutrient content of selected dietary mixture	NAS–NRC estimated requirements (1994)
Yellow maize	52.8	Crude protein 179g/kg	180g/kg
Oat meal	8.9	ME 2729kcal/kg	2880kcal/kg
Wheat bran	21.3	Calcium 13g/kg	9.0g/kg
Fishmeal	11.4	Phosphorus 11g/kg	4.0 non-phytate P
Bone meal	2.9		
Dried skim milk	2.1		
Oyster shell	0.6		

TABLE 4.28 Dietary self-selection by chicks and effect on nutrient intake (from Dove, 1935).

Age	Maize meal	Fish meal	Bone meal	Oyster shell flour	Wheat bran	Wheat middlings	Oat meal
	% of diet						
0–3.5 d	29.5	6.3	3.0	0.7	12.6	14.4	33.6
3–7 d	32.5	7.9	4.4	0.9	12.8	10.3	31.0
7–14 d	30.6	7.7	3.8	0.7	18.3	4.7	34.2
2–3 w	47.4	8.6	2.4	0.1	11.9	4.0	25.6
3–5 w	57.3	7.9	2.3	1.1	10.1	2.3	19.0
5–7 w	61.9	9.4	2.2	0.7	12.6	2.2	11.0
7–9 w	65.4	10.2	2.9	0.8	11.7	3.2	5.8
9–11 w	67.1	11.5	3.6	0.9	9.7	1.7	5.5
11–13 w	68.9	13.1	4.1	1.6	8.0	1.4	2.9

TABLE 4.29 Percentage of dietary ingredients self-selected by Rhode Island Red chicks over the period from hatching to 13 weeks of age (from Dove, 1935).

of 'nutritional wisdom', giving them the ability to select from the various feeds on offer and construct their own diet according to their actual needs and production capacity. It is known that the wild ancestors of modern chickens have this ability and there is strong evidence that at least some types of modern poultry also possess it. Table 4.28 shows that chicks offered a choice of feeds took in a balance of nutrients that was close to the NAS–NRC (1994) estimates of requirements.

One question that the data of Dove

(1935) helps to answer is at what age poultry can be introduced to choice feeding. Recent data on this important issue are lacking.

Dove (1935) used a total of 3,000 Rhode Island Red chicks in his study and allowed them a choice of maize meal, fishmeal, bone meal, oyster shell flour, wheat bran, wheat middlings and oat meal over the period from hatching to 13 weeks of age. The percentage of each feedstuff in the total intake was then measured. The results are shown in Table 4.29.

These data indicated a fairly stable intake

of the feedstuffs initially, followed by an increasing intake of maize meal and fish-meal and a marked decrease in the intake of wheat bran, wheat middlings and oat meal after 2 weeks. The intakes of bone meal and oyster shell flour were more consistent.

The data suggest that chickens are able to discriminate between different feedstuffs after 2–3 weeks, selecting the feedstuffs that provide their nutritive requirements.

According to one expert on whole grain feeding (Bennett, 2015), whole grain can be fed from an early age and the diet can contain 50% whole grain by 3 weeks of age.

This researcher summarised the effects of whole grain feeding in laying, broiler and turkey stock as follows:

1. In the first 6 weeks of feeding whole grain, the gizzard responds by increasing muscle mass. During this adjustment period, growth slows and feed:gain ratio increases. For every 1% of whole grain added to the diet in the first 6 weeks of age, live weight gain declines by 0.09% and feed:gain ratio increases by 0.07%.
2. The muscle of the gizzard does not increase noticeably until the birds are fed 20% or more whole grain.
3. Delaying the age at which whole grain is fed does not help the gizzard to adapt to grinding the grain.
4. Feeding grit does not have a significant effect on how efficiently birds use whole grain.
5. After the birds have been fed the maximum level of whole grain for 3 weeks, the gizzard will grind whole grain as efficiently as a hammer mill or pellet mill. At this stage, feeding whole grain will not alter weight gain or feed efficiency.
6. Chickens and turkeys react very similarly to being fed whole wheat or barley. The only significant difference is that initial growth rate is slowed more in turkeys fed whole barley than in turkeys fed whole wheat or chickens fed either type of whole grain.
7. Feeding whole grain reduces leg and skeletal problems. The reduction is most significant in feeding programmes which slow growth in the first 6 weeks of age but is still noticeable in programmes where birds fed whole grain grow as quickly as those fed pellets. Both chickens and turkeys benefit.
8. The only consistent effect on carcass yield is associated with the 20 or 30g increase in gizzard weight in birds fed whole grain.

Laying Hens and Dual-Purpose Chickens

A suggested approach to applying these findings with organic laying or dual-purpose chickens is to start the birds on starter feed in mash or crumbled form for the first 2 weeks from hatching. The presence of a mother hen will help. Using double the usual number of feeders is recommended. After 2 weeks it is suggested that some whole wheat or kibbled maize be sprinkled on top of the starter feed in one half of the feeders. Once the birds start to eat substantial amounts of whole grain the feed in half of the feeders should gradually be replaced with whole grain and the feed in the other half should be gradually blended with a supplement so that by the time the birds are consuming 100% whole grain in one feeder they should be consuming 100% supplement from the other feeder. Thereafter the birds should be well adapted to a choice feeding regime.

At around 8 weeks of age the birds can be switched over gradually from a starter supplement to a pullet developer supplement and continue on this regime up to about 18 weeks of age. At this time the pullet

Ingredient, kg	Starter 0–8 w	Pullet grower 8–18 w	Laying/breeding hens
Barley, ground	180	90	–
Wheat, ground	42	–	290
Wheatfeed	200	390	166
Oats, ground	–	100	–
Soya bean meal	300	–	350
Canola meal	–	100	–
Field peas, ground	200	200	50
Dehydrated lucerne or grass meal	–	50	–
Ground limestone	28	26	100
Dicalcium phosphate	24	18	18
Salt (NaCl)	6	6	6
Vitamin Premix	10	10	10
Trace mineral Premix	10	10	10
Total, kg	1000	1000	1000

TABLE 4.30 Examples of supplement mixtures (air-dry basis) for use in choice feeding of organic laying stock with access to whole grain*.

*Regular grit of a suitable granular size (to aid grinding in the gizzard) should also be available. Calcareous grit such as oyster shell grit needs also to be provided for laying hens.

Reference	Conventional diet g/hen/day	Choice feeding g/hen/day	Saving in feed intake %
Henuk et al. (2000b)*	123.8	120.3	2.9
Blair et al. (1973)*	116.2	108.9	6.7
Leeson and Summers (1978)*	114.4	107.2	6.7
Leeson and Summers (1979)*	118.4	110.7	7.0
Henuk et al. (2000a)*	126.5	114.6	10.4
Karunajeewa (1978)*	132.5	118.5	11.8

TABLE 4.31 Estimated feed savings by adopting choice feeding with laying hens (from Blair, 2008).

*Cited by Blair (2008).

developer supplement should gradually be replaced with a layer supplement and a third set of feeders containing calcareous grit (limestone or oyster shell) should be provided. The purpose in providing calcareous grit is to build up the calcium reserves in the hens, in preparation for egg laying.

Examples of supplement mixtures that can be used with laying chickens are shown in Table 4.30.

Choice feeding should work well with birds provided access to forage since it allows the birds to regulate the intake of energy and nutrients according to the nutrients supplied by foraging.

Results from several studies on the effect

of choice feeding on performance of laying hens suggest that when birds are offered a free choice of feeds they consume less feed in total than birds receiving a conventional complete diet (Table 4.31).

Bennett (2006b) made the following recommendations for small-scale egg producers:

1. Do not give the hens too many choices. Hens can handle up to three choices quite well (grain, supplement and limestone or oyster shell). When using more than one grain, such as wheat and barley, mix them together in the same feeder.

2. Give the hens choices that are nutritionally distinct. For example, grain is high in starch and energy, supplement is high in protein and vitamins, and limestone is high in calcium. When provided with such clear choices, the hens learn which feeders to go to and how much to eat in order to meet their basic nutritional needs. Some choices may not be clear enough for the hens. For example, wheat and peas both are high in starch and have moderate levels of protein. Having separate feeders containing wheat and peas may not provide a distinct enough nutritional difference for the birds to detect.

3. Introduce the whole grain and choice feeding a month before the start of egg production, i.e. at about 15 weeks of age (when the birds have not been introduced to choice feeding earlier). This adjustment period will allow the birds time to learn how to choice feed themselves before they are exposed to the nutritional demands of egg production. It will also allow the pullets the opportunity to increase their calcium intake and build up the calcium reserves in their bones before they start to lay eggs. Finally, it takes the gizzard 3 weeks to build muscle mass and the intent is for the hens to be

able to grind the grain efficiently in this organ once egg production begins.

4. Do not feed vitamins or trace minerals in a separate feeder. Use the supplement as a source of these nutrients. If vitamins or trace minerals are placed in a separate feeder, some birds may not eat them because they do not like the taste while other birds may over-consume them and suffer toxic side-effects.

5. Give the birds adequate feeder space. With a large flock, several feeders are required for each feed. For a 100-hen flock, 2 hanging feeders each of grain, supplement and limestone are suggested.

6. Purchase a supplement designed to be mixed with grain or grain and limestone (or oyster shell) to provide a complete laying hen diet. A supplement formulated in this way will contain 250–400g crude protein per kg protein. A grower supplement may be used prior to the start of egg production but a layer supplement should be used once the birds start to lay.

The birds will readily consume whole wheat, whole oats or whole barley, but they have difficulty with whole maize, which needs to be kibbled (reduced in particle size). Hens can successfully consume 70% of their diet as whole grain when it is choice-fed. It is important to note that when whole grain, supplement and limestone are mixed together into a traditional laying hen diet and offered in one feeder, the whole grain should not comprise more than 50% of the diet. The rest of the grain in the feed should be ground. At higher levels of whole grain, the hens sometimes have trouble finding the supplement in the feed mixture. When the grain, supplement and limestone are in different feeders, these separation problems are avoided.

Research has shown that choice feeding can be used successfully with growing pullets and growing broilers. Egg-type chickens are known to adapt more quickly than meat-type, and industry findings suggest that unimproved chickens adapt more quickly than modern hybrids. Brown egg layers have been shown to adapt more quickly than white or tinted egg layers.

Experience and group learning are important. Growing chickens take about 10 days to learn to balance the intakes of whole grain and protein supplement accurately. They need to be in groups of at least eight birds and be offered the protein concentrate (in mash or crumble form) and whole grain in identical, adjacent troughs, or in the same feed trough.

Meat-Type Chickens, Roasters

Fanatico et al. (2013) found that free choice feeding worked well with meat-type chickens under free range conditions. In their study the birds all received a combined starter/grower diet during the brooding period (0–27 days). During the grower period (28–49 d) one group received the same feed only while another group received the same diet along with free choice ingredients (cracked maize, whole wheat, soya bean meal, fishmeal, crushed oyster shell, kelp meal, bone meal, and trace mineral salt: all ingredients provided in separate feeders) so that the birds could learn to self-select. During the finisher period (49–83 d), the second group received only free choice ingredients. Insoluble grit was provided to all birds to help grind whole grains and forage. Feed and water were provided both indoors and outdoors.

Analysis of the intakes of the major feeds selected showed that amounts of maize decreased, wheat increased, soya bean meal decreased sharply and fishmeal increased initially in the diet during the growth period and remained the same in the finisher period.

The formulated diet was a commercial product (200g/kg protein), whereas the free choice diet chosen by birds was much lower in protein (130g/kg). Final live weights did not differ between treatments. However, ready to cook yield and breast meat yield were higher in the birds fed the complete feed, most likely due to amino acid supplementation. The diet chosen by free choice birds was less expensive than the complete feed.

These results suggest that meat-type chickens can be choice-fed using whole wheat after about 4 weeks of age. The birds should be well feathered and the weather favourable before they are allowed outdoors to forage.

An approach similar to that recommended for layer stock is suggested, namely to start the birds on broiler starter feed in mash or crumbled form for the first 2 weeks from hatching. The presence of a mother

Ingredient, kg	Growth period 4–12 weeks
Wheat, ground	300
Wheatfeed	150
Soya bean meal	470
Ground limestone	24
Dicalcium phosphate	30
Salt (NaCl)	6
Vitamin premix	10
Trace mineral Premix	10
Total, kg	1000

TABLE 4.32 Example of a broiler supplement mixture (air-dry basis) for use in choice feeding of organic meat-type chickens with access to whole grain.*

*Regular grit of a suitable granular size (to aid grinding in the gizzard) should also be available.

hen will help. Using double the usual number of feeders is recommended. After 2 weeks it is suggested that some whole wheat be sprinkled on top of the starter feed in one half of the feeders. Once the birds start to eat substantial amounts of whole grain, the feed in half of the feeders should gradually be replaced with whole grain and the feed in the other half of the feeders should be gradually blended with a broiler grower supplement (Table 4.32) so that by the time the birds are consuming 100% whole grain in one feeder they should be consuming 100% supplement from the other feeder. Thereafter the birds should be well adapted to a choice feeding regime.

Turkeys

Choice feeding has been shown to be successful with turkey poults started on choice feeding at 7 days of age, productivity being slightly better when wheat was the whole grain offered. A feeding regime similar to that described above for growing chickens and using supplements designed for use with the particular strain being used is recommended. Choice feeding has also been shown to be successful with turkey hens.

Turkeys grow well on range when environmental conditions are favourable, especially heritage strains.

Ducks

Choice feeding cannot yet be recommended for ducks and other waterfowl due to a lack of supporting evidence on the use of whole grain in combination with appropriate supplements. Work in Vietnam (cited by Blair, 2008) in which commercial diets with the same energy level but differing levels of crude protein were fed to growing meat-type ducklings *ad libitum*, found that ducklings preferred high-protein to low-protein

feeds. This resulted in excess protein intake and higher protein conversion ratios. This study concluded that choice feeding of the type used was not an economically viable system for growing meat-type ducks.

FORAGE

Three aspects of forage feeding are of interest:

a) How much do free-range poultry consume?
b) How well is it utilized by poultry?
c) What is its significance in relation to nutrient needs?

Another consideration is that one of the most important egg quality parameters for the consumer is yolk colour, which can be affected by forage intake and quality.

Work from Denmark has shown that modern hybrid laying hens do well when fed whole wheat and have access to good pasture (Horsted et al., 2007). The hens were moved regularly in a rotation between different forage crops for a period of 130 days, half being fed organic layer feed and half being fed whole wheat only. The forage crops were grass/clover, pea/vetch/oats, lupin and quinoa. After a period of 6 to 7 weeks the intake of wheat increased to approximately 100g per hen per day and egg production increased to the same level as for the hens fed complete layer feed. Egg weight and body weight increase were similar in both groups of hens. Crop analysis showed that wheat-fed hens ate fewer seeds from the pasture, but more plant material, oyster shells, insoluble grit and soil. The incidence of floor eggs was significantly higher in the hens fed layer feed, whereas wheat-fed hens only rarely laid floor eggs (no explanation). Irrespective of treatment, hens were found to have excellent health and welfare. This work showed that nutrient-restricted, high-producing organic layers are capable of finding and utilising considerable amounts of different feed items from a cultivated foraging area without negative effects on their productivity, health and welfare. These conclusions are supported by findings of other researchers.

One of the results of access to soil and forage is that birds ingest arthropods such as insects and earthworms, which can provide an additional source of nutrients. Intake of insects by jungle fowl chicks and wild turkey poults is known to possibly exceed 50% of their diet, and adult females increase their intake of insects at the time of reproduction. Research findings reviewed by Ravindran and Blair (1993) suggest that the type of insect preferred by poultry is soft bodied.

Research findings cited by Blair (2008) indicate that earthworms can provide a useful amount of protein for poultry. The concerns are that the earthworms may concentrate heavy metals and contaminants present in the soil and may act as intermediate hosts for cestode worms and disease vectors such as that causing blackhead in turkeys. This concern about disease spread is minimized in some tropical countries by collecting the earthworms and sun-drying them before feeding to poultry.

The significance of the above results for organic producers is that poultry allowed to range may be able to obtain a substantial proportion of their nutrient needs from pasture plants, insects and earthworms, etc. However, the intake is difficult to quantify, therefore the most appropriate approach is to choice feed the flock on grain and supplement. By feeding in this way the birds can then adjust their intakes of protein and energy according to the amounts obtained from foraging.

BREEDS FOR ORGANIC PRODUCTION

Organic production is largely consumer driven, therefore it is important to take into account consumer attitudes in selecting the appropriate breeds and strains for organic poultry production. A yellow-skinned bird is preferred by many consumers, others preferring white-skinned birds or birds with coloured skin. Some consumers prefer white-shelled eggs, others preferring eggs with tinted or coloured shells. Most consumers appear to prefer eggs with highly pigmented yolks.

There are four major sectors of poultry produce: eggs; chicken meat; turkey meat; and niche meat and egg products such as game birds, waterfowl (ducks and geese), ratites (ostriches and emus), squab (pigeons), silkie chickens, quail and quail eggs and game birds (pheasants, partridges and tinamou). All are being produced organically in various parts of the world, the sectors differing in economic value in different regions.

Laying Hens

For brown egg production, the genetic strains have been developed mainly from a cross of Rhode Island Reds and Barred Plymouth Rocks. Laying breeds that are suitable for free-range and organic egg production include the hybrids ISA Brown and Hyline and traditional breeds such as Light Sussex and Rhode Island Red. Other breeds popular in some organic systems are Black Rocks and Hebden Blacks. A newer breed, Bovan Goldline, has a reputation of being docile and exhibiting less of a trait that can affect the choice of breed for organic egg production – feather pecking – which can lead to cannibalism.

Scandinavian research has shown that breeds such as the Skalborg and the hybrid SLU-1329 (Rhode Island Red × White Leghorn) are more suitable for organic production systems than conventional hybrids.

According to research findings in Europe, ISA Brown under organic experimental conditions had higher egg production than New Hampshires and White Leghorns but showed a higher level of cannibalism. In other work, Hyline White and Hyline Brown laid more eggs than two traditional Italian breeds (Ermellinata di Rovigo and Robusta Maculata) under organic conditions.

Some organic egg producers keep a flock for several years and run multiple ages together. The introduction of a different coloured breed each year makes it possible to keep track of the age of each of the hens in these flocks.

Organic producers, therefore, need to get advice from local agencies on which breeds of laying hens are best suited for egg production in that particular region, based not only on their inherent egg-laying ability but also on their suitability for outdoor and pasture-based based production and on the type of market being served. Other attributes that should also be taken into account are the brooding and mothering abilities of the breeds available since many organic farmers like to breed and brood their birds naturally; also their suitability for continued production over several seasons.

Meat Chickens

The stipulation that the market age of organic meat chickens must be a minimum of around 80 days requires that breeds and strains that are more slow-growing than the modern hybrids used in conventional broiler production have to be used. Otherwise the birds are likely to be too heavy when marketed, typically reaching market weight at 5–6 weeks, and may suffer from musculoskeletal problems.

Many slow-growing genotypes are available in Europe, and researchers have shown that although the growth rate of these genotypes is lower than that of fast-growing birds, they are more adapted to natural systems, have fewer health issues and the quality of their meat is more appropriate for a specialty or gourmet market.

Slow-growing birds are a key in Label Rouge production, taking 12 weeks to reach market weight. As a result, the meat has a stronger flavour. The carcass is generally more elongated, with a smaller breast and larger legs than conventional carcasses. The types of birds used are more suited to outdoor production than Cornish crosses.

The slower growing breeds are less efficient in feed utilisation than modern hybrid broiler and roaster strains, but generally provide a higher profit margin to producers due to the premium paid for the meat in appropriate markets.

In Europe, the slower growing birds are mainly supplied by poultry breeding companies. Examples of suitable slower growing breeds include: Sasso, Poulet Bronze, Poulet Grey, Sherwood Gold, Hubbard ISA 657, 257, PAC57, Light Sussex, White Sussex, Ixworth, White Dorking, Redbro, MasterGris, JA 57, Colopak, Co'Nu and Gris Barre.

In general, active and inquisitive breeds should be chosen as they are likely to be better rangers and foragers than the less active broiler hybrids. Breeds that exhibit extreme escape responses and peck the ground in a stereotypical manner should be avoided as they will be more difficult to manage and likely to engage in aggressive feather pecking and cannibalism. Coloured feathering helps to reduce the risk of predation, especially in woodland situations, whereas white feathering is likely to increase the risk.

Dual-Purpose Breeds

For many producers the ideal chicken for organic production is one of the dual-purpose breeds, i.e. one developed for both egg and meat production. These breeds fit into organic poultry systems more appropriately than breeds developed specifically for egg or meat production. Many are heritage breeds and have a much lower incidence of health problems such as ascites (water-belly) and sudden death syndrome (SDS) than found with commercial meat breeds. One disadvantage is that the meat from dual-purpose breeds is not available until the end of the laying period when the hens are replaced. Surplus cockerels, however, can be marketed earlier.

Among dual-purpose breeds are the Rhode Island Red, Barred Plymouth Rock, New Hampshire × Barred Rock, Rhode Island Red × Columbian Rock, and the Shaver Red Sex-Link and Harco Black Sex-Link hybrids. The Favorelle is a white-skinned chicken that was developed in France as a dual-purpose breed.

Turkeys

Wild turkeys still exist in many parts of the US. While they are the precursors of the modern turkey, a high proportion of the meat is considered dark, including the breast. Heritage turkeys grow more slowly than commercial birds and the meat has a more highly developed flavour, especially when they are allowed to forage on pasture. They are typically slaughtered at 7–8 months of age, whereas commercial turkeys are ready for market in 3–4 months. Heritage turkeys are single-breasted rather than double-breasted like the commercial

varieties, and consequently have less white meat. A typical Large White has nearly 70% white meat whereas the heritage breeds have about 50–50 white to dark meat.

In North America most consumers prefer the breast, or white meat of a turkey. As a result, many organic turkey producers continue to use the commercial strains of turkeys and use low-energy diets to reduce their growth rate.

Heritage turkeys adapt well to organic production systems since they are more disease resistant and are good foragers. In addition, they are strong flyers, and can mate naturally and raise their young successfully. Several heritage turkey strains are available and the preference varies from region to region. The strains are usually distinguished according to their colour and region of origin. They include Standard Bronze, Narragansett, Bourbon Red, Jersey Buff, Slate, White Holland, Black Spanish, Beltsville Small White and Royal Palm. The Narragansett is one of the oldest varieties available and once served as the foundation of the New England turkey industry. Royal Palm (also known as Crollweitzer or Pied) was also popular prior to the era of commercial turkey production. Both varieties were traditionally grown on family farms.

Waterfowl

The choice of these species depends on their intended use (i.e. meat, eggs, weeding, herding, or guard animals).

Ducks are primarily kept for meat because of their rapid growth, hardiness and ease of handling. Rapid growth combined with good egg production make the Pekin the most popular breed of duck for meat. Pekins are capable of growing rapidly and reaching 3.2kg in 7 weeks. Other good meat strains are the Aylesbury, Rouen and Muscovy.

The breeds for egg production include the Indian Runner and Khaki-Campbell. These ducks will lay as well as a good strain of White Leghorn. The Tsaiya, a native duck in Taiwan, is used for egg production. In Indonesia the Alabio and Bali breeds are common while in China the native Maya is used.

Some duck breeds are considered multi-purpose in that they produce a large number of eggs but also have a meatier carcass than most egg-laying breeds. These include the Aylesbury, which is used in many of the organic duck farms in the UK, the Cayuga, and the Maya (China).

The goose is one of the most ancient of domesticated birds now farmed, with the greatest concentration found in Asia. An important feature of geese is their ability to consume green forages and crop residues.

Geese are divided into three categories – light, medium and heavy. The most common geese raised for meat are in the heavy category and include the Toulouse, Embden, African, and Pilgrim. The Toulouse and Embden breeds are most popular in North America.

Chinese geese are commonly raised as weeder geese and fit well into some organic production systems for the biological control of weeds.

Quail

Two types of quail have been domesticated and raised as food animals, Japanese and Bobwhite quail.

Japanese quail (*Coturnix japonica*) are native to Asia. Breeding programmes have developed lines of Japanese quail specific for egg and meat production. Japanese quail production for meat occurs primarily in Europe while production of quail eggs occurs primarily in Asia and South America.

Bobwhite quail (*Colinus virginianus*) are native to the US and are primarily raised for slaughter and sale as quail meat or for release in hunting areas. The varieties differ in body size, the smaller varieties tending to lay more eggs than the larger varieties.

Ostriches and Emus

The emu (*Dromaius novaehollandiae*) and ostrich (*Struthio camelus*) are ratites, i.e. flightless birds with broad, rounded breast plates missing the keel to which the breast or flight muscles attach. Both are now being farmed in several countries.

Ostriches are native to South Africa, where they have been raised commercially for more than 100 years. Ostriches breed between 3 and 4 years of age. Chicks reach maturity within 6 months, adults weighing 95–175kg and measuring 2–3m in height. Thus, they require careful handling. Ostriches are farmed mainly for meat, hides (leather), and feathers.

Emus are native to Australia. The US first imported emus between 1930 and 1950 but commercial emu farming in North America did not begin until the late 1980s. The female emu begins to breed between 18 months and 3 years of age, and may continue to produce eggs for more than 15 years. The emu grows to full size within 2 years, when it is 1.5–1.8m tall and weighs up to 65–70kg. Emu products include leather, meat, and decorative egg shells.

NUTRITION AND HEALTH

Disease problems in organic flocks need to be minimized since they have an adverse effect on the birds, and the eggs and meat produced may pose problems for the human consumer.

Disease prevention in organic farming is

FIGURE 4.3 Ranged poultry.

based on the principle that an animal that is allowed to exhibit natural behaviour, is not subject to stress and is fed an optimal diet will have a greater ability to cope with infections and health problems than animals reared in a conventional way. Fewer medical treatments are then necessary. Good husbandry practices, including appropriate choice of breed and strains, housing conditions, space allowance, sanitation practices and prompt treatment of diseases, will result in a high level of animal health. Feeding the animals well is a first step since this helps the birds to develop an active immunity to diseases.

Raising young chicks with their mothers has several advantages, including the development of immunity to some diseases through the ingestion of droppings from the hen while learning to forage.

Zinc is an important element in fighting infections and is sometimes used for disease control in conventional production. Use of this trace mineral is not approved for that purpose in organic production, therefore producers are advised to use phytase in their dietary formulations to help ensure that the maximal amount of dietary zinc is available to the animal and not bound in the dietary ingredients with phytate.

Strict biosecurity is important in any poultry operation and particularly in organic operations. Wild birds, particularly waterfowl, can carry diseases such as avian influenza that harm domestic poultry, therefore it is important to exclude wild waterfowl from free-range poultry areas. Outdoor feeders should be designed and sited so that they do not attract wild birds. If necessary, netting can be placed over outdoor yards.

Other measures to avoid or reduce

disease risks include (where possible) the use of an all-in, all-out management system, based on the principle that depopulation at the end of a flock reduces the pathogen load because some pathogens die when there is no host. Mixing ages in a flock can pose a risk in that older birds may be carriers of disease for younger birds. Likewise, mixing species can result in some species carrying diseases to other species, e.g. domestic ducks and geese can carry diseases that infect chickens. The benefits of introducing older birds (or other species) to teach newly hatched chicks or poults to eat have therefore to be weighed against the risks.

Providing birds with access to outdoors has the advantages of providing exercise and fresh air but has the disadvantages of exposing the birds to predators and disease threats in soil and water, and from wild birds and other animals in the environment. Predators (hawks, foxes and martens) are known to be a cause of losses in free-range systems. Another part of the programme should be to ensure that the stock is not exposed to diseases spread by rodents and wild birds. Appropriate housing and yard design should be adopted so that these threats are minimized.

The occurrence of organisms causing enteric diseases in humans – such as salmonella – has been shown to be lower in organic poultry than in conventional poultry. However, home-processed chicken meat from organic farms is more likely to be contaminated than commercially-processed chicken meat, indicating where preventative measures have to be taken.

Conventional poultry are usually vaccinated against a variety of diseases including Marek's disease, Newcastle disease, infectious bronchitis, infectious laryngotracheitis, fowl pox and fowl cholera. These diseases can also affect organic poultry, therefore a routine vaccination programme is recommended, where allowed under the local organic regulations.

A main health problem in poultry is gastro-intestinal disease. Relevant approaches to this problem include the enhancement of immunity, using whole grains in the diet to encourage gizzard development, and including some fibrous ingredients to the diet to encourage fermentation in the large intestine and stimulate the growth of beneficial bifidobacteria.

Probiotics are a potential tool for reducing intestinal contamination with disease-causing and foodborne bacteria. They may also be useful in the prevention or treatment of coccidiosis.

Prebiotics are indigestible or low-digestible feed ingredients that benefit the host organism by selectively stimulating the growth or activity of beneficial bacteria in the hind-gut. Part of the reasoning for the use of dietary fibrous sources that ferment in the large intestine is that they may produce butyrate, a short-chain fatty acid (SCFA). Butyrate and other SCFAs are important in relation to the absorption of electrolytes by the large intestine and may play a role in preventing certain types of diarrhoea (and cancer in humans). This group includes chicory and Jerusalem artichoke, which contain inulin-type fructans in the sap and roots, also *Cichorium intybus* (compositeae) a natural plant source of fructooligosaccharides.

Some herbs such as allium, thymus, anthriscus and ferule are also known to stimulate acid production by lactobacilli and might be useful prebiotics for use in poultry (and human) nutrition.

One of the disadvantages of using diets containing some partly indigestible carbohydrates is that they can lead to increased worm infestations because they provide more of a nutrient source in the lower gut. Therefore, poultry producers should use highly digestible diets during outbreaks of worm (helminth) infestations and, where possible, should use liquid whey as a dietary supplement. This product is known to be useful in helping to control ascarid infestations. A good grazing management plan should also be used, with rotation of pastures and yards. Most helminths are strictly host specific and mixed grazing is known to be useful in helminth control.

The ascarid *Heterakis gallinarum* is a roundworm of approximately 1–2cm in length, found mainly in the caeca of many bird species, including chickens and turkeys. Although this worm does not present a serious disease problem, *H. gallinarum* is probably the most important of the poultry nematodes because of the important role it plays in the epidemiology of histomoniasis (blackhead). The egg provides the most natural route of transmission for the protozoan *Histomonas meleagridis*, the causative organism for blackhead. As with all helminths of poultry, this is a particular risk associated with free-ranging and organic poultry systems. A Danish survey has shown that 72.5% of organic/free range birds were infected with *Heterakis gallinarum*. The control of *H. gallinarum* in free-range flocks should focus on maintaining the ranging area at low levels of parasite infestation. In most situations this is best achieved by regular rotation of the ranging paddocks. Completion of the life cycle of the parasite can be kept low by using the same range only every third year.

As noted in Chapter 3, supplementation of the diet with seaweed has been shown to be effective in the control of parasites in pigs, markedly reducing the incidence of liver condemnations from ascarid damage in pigs at slaughter. This aspect of seaweed inclusion is of interest to organic farmers but does not appear to have been investigated in poultry.

Another approach to improving the gut function is the replacement of disease organisms in the gut with beneficial organisms (competitive exclusion). This allows the intestinal epithelium and host microflora to act as natural barriers to damage from pathogenic bacteria, antigens, and toxic substances inside the gut. Products approved for use in organic production are mixed cultures derived mainly from the caecal contents and/or gut wall of domestic birds. The treatment is normally given to newly hatched chicks or turkey poults as soon as possible after hatching, either by spraying at the hatchery or at the farm, or by addition to the first drinking water. Organic farmers should obtain veterinary advice on whether this procedure is appropriate for their farm, since the young chicks may have been hatched by broody hens and have ingested droppings from the hen while learning to forage.

Some products of this type have been shown to reduce salmonella in poultry.

Several research findings have demonstrated the beneficial effects of whole grain feeding on the gut microflora and the overall health of poultry. They show that a better developed gizzard has an important function as a barrier organ in preventing pathogenic bacteria from entering the distal digestive tract. For instance, it has been shown that broilers infected with an antibiotic-resistant strain of *Salmonella typhimurium* at 15 days of age had lower numbers of the organism in the gizzard and ileum when fed a diet containing whole wheat. As a result of such studies some researchers have suggested that whole grain could be considered as an effective alternative to antibiotic growth promoters.

Birds fed a diet containing whole wheat have been shown to have a significantly lower (2.5 times) oocyst output from coccidiosis infection than birds fed on a diet with ground wheat. This suggests that an active functioning gizzard can play a role in resistance to coccidiosis. Other research showed an increase in intestinal counts of some beneficial Lactobacillus species, a decrease in *E. coli* counts, and a lower number of pathogens such as salmonella spp. or *Clostridium perfringens* (responsible for necrotic enteritis) as a result of whole grain feeding. Whole grain incorporated into pelleted diets at 200g/kg has been shown to reduce the incidence of proventricular dilatation and mortality from ascites in broilers.

Another interesting finding was that feed containing flaxseed as a source of n-3 fatty acids was beneficial in reducing lesions caused by coccidiosis. These findings are of potential interest to producers of 'designer' eggs, obtained from hens fed diets containing flaxseed.

As noted above, the use of slow-growing strains reduces the incidence of skeletal lesions in organic broilers. There is also evidence that some native strains of laying chickens have a greater resistance to infection from intestinal worms than modern hybrids.

REFERENCES

ARC (1975). Agricultural Research Council: The Nutrient Requirements of Farm Livestock, No. 1 Poultry. Agricultural Research Council, London, UK.

Bennett, C. (2006a). Organic Diets for Small Flocks. Publication, Manitoba Agricure. http://www.gov.mb.ca/agriculture/livestock/poultry/bba01s20.html, accessed November 2, 2015.

Bennett, C. (2006b) Choice-Feeding of Small Laying Hen Flocks. Extension Report, Manitoba Agriculture, Food and Rural Initiatives, Winnipeg, Canada, pp 1–2.

Bennett, C. (2015). Personal communication dated November 30, 2015.

Blair, R. (2008). Nutrition and Feeding of Organic Poultry. CAB International, Wallingford, Oxford, UK. 314 pp.

Blair, R. (2012). Organic Production and Food Quality: A Down to Earth Analysis. Wiley-Blackwell, Hoboken, NJ 07030, USA, 296 pp.

Blair, R., Dewar, W.A. and Downie, J.N. (1973). Egg production responses of hens given a complete mash or unground grain together with concentrate pellets. British Poultry Science 14, 373–377.

Bolton, W. and Blair, R. (1974). Poultry Nutrition. Bulletin 174, Ministry of Agriculture, Fisheries and Food. London: HSMO, 135 pp.

Dove, F.W. (1935) A study of individuality in the nutritive instincts and of the causes and effects of variations in the selection of food. American Naturalist 69 (Suppl.), 469–543.

Fanatico, A.C., Owens, C.M. and. Emmert, J.L. (2009). Organic poultry production in the United States: Broilers. Journal of Applied Poultry Research 18:355–366. (Available online at: http://dx.doi.org/10.3382/japr.2008-00123 (accessed August 14, 2015).

Fanatico, A.C., Brewer, V.B., Owens-Henning, C.M., Donoghue, D.J. and Donoghue, A.M. (2013). Free-choice feeding of free-range meat chickens. Journal of Applied Poultry Research 22: 750–758.

Horsted, K., Hermansen, J.E. and Ranvig, H. (2007). Crop content in nutrient-restricted versus non-restricted organic laying hens with access to different forage vegetations. British Poultry Science 48: 177–184.

Lampkin, N. (1997). Organic Poultry Production, Final report to MAFF 1997. Welsh Institute of Rural Studies, UK.

Lewis, P.D., Perry, G.C., Farmer, L.J., Patterson, R.L.S. (1997). Responses of two genotypes of chicken to the diets and stocking densities typical of UK and 'Label Rouge' production systems: I. Performance, behaviour and carcass composition. Meat Science 45: 501–516.

NAS–NRC (1994). Nutrient Requirements of Poultry, Ninth Revised edn. National Research

Council, National Academy of Sciences, Washington, DC.

Ravindran, V. and Blair, R. (1993). Feed resources for poultry production in Asia and the Pacific. III Animal protein sources. World's Poultry Science Journal 49: 219–235.

Scheideler, S.E. and Sell, J.L. (1997). Nutrition Guidelines for Ostriches and Emus. Publication PM-1696, Extension Division, Iowa State University, Ames, Iowa, pp. 1–4.

Steenfeldt, S., Engberg, R.M. and Kjær, J.B. (2001). Feeding roughage to laying hens affects egg production, gastrointestinal parameters and mortality. Proceedings of 13th European Symposium on Poultry Nutrition. Blankeberge, Belgium, pp. 238–239.

Ullrey, D.E. and Allen, M.E. (1996). Nutrition and feeding of ostriches. Animal Feed Science and Technology 59, 27–36.

Ingredient	Maximum inclusion rate (g/kg diet)*		
	Starting	Growing/Finishing	Laying/Breeding
Barley	150	300	400
Cassava	50	100	150
Canola meal	50	150	50
Canola, full-fat	50	100	0
Cottonseed meal	100	150	150
Faba beans	100	200	200
Fishmeal, white	100	50	50
Grass meal	50	100	50
Herring meal	50	25	25
Lucerne meal, dehydrated	0	50	50
Kelp (seaweed) meal	0	25	25
Lupin seed	50	100	150
Maize	600	500	600
Molasses	20	40	20
Oats	10	20	20
Peanut (groundnut) meal	100	150	250
Peas	75	300	200
Rice bran	50	150	250
Rye	0	25	0
Safflower meal	25	150	100
Sesame meal	25	150	100
Sorghum (milo)	400	400	500
Soya bean meal	400	250	400
Soya beans, full-fat, heat-treated	100	200	200
Sunflower meal	0	200	100
Triticale	25	150	100
Wheat	600	500	600
Wheat middlings	100	400	600
Yeast, brewer's dried	50	50	50

APPENDIX TABLE 4.1 Recommended maximum inclusion rates of feedstuffs in organic poultry diets.

* The levels shown above assume the dietary inclusion of a supplementary enzyme mixture (to ensure optimum utilization).

CHAPTER 5

Feeding Organic Cattle and Other Ruminants

Cattle and other ruminants have evolved to subsist mainly on forage and browse. Therefore forages either grazed or fed in conserved form may provide all the nutrients required. However, forages may not be available in sufficient quantity at all times during the year; also, the quality may be low. As a result, ruminant animals, especially dairy cows, are likely to need supplementary feeding periodically. The supplements and additives that are permitted are detailed in Chapter 2.

In general the regulations require that organic cattle, sheep, and goats must have free access to organic pasture for the entire grazing season. Depending on local environmental conditions (e.g., temperature, rainfall), the grazing season will range from 120 to 365 days per year. This period is specific to the farm's geographic climate, but must be at least 120 days. Owing to weather, season, or climate, the grazing season may or may not be continuous.

The diets must contain at least 30% dry matter (on average) from certified organic pasture. Dry matter intake is the amount of feed an animal consumes per day on a moisture-free basis. The remainder of the diet must also be certified organic, and includes hay, grain, and other agricultural products.

Outside of the grazing season, organically raised ruminants must have free access to the outdoors all year round except under specified conditions (e.g. inclement weather). Cattle slaughter stock for the last fifth of their lives (up to 120 days) are exempt from the requirement that 30% dry matter intake must be from pasture.

Feeding ruminant animals is more complex than the feeding of pigs and poultry, therefore an understanding of the principles of ruminant nutrition is necessary before appropriate feeding programmes can be outlined.

DIGESTION AND ABSORPTION OF NUTRIENTS

The following summary outline of digestion and absorption in ruminant animals provides a basic understanding of how the feed is digested and the nutrients absorbed. A more detailed description is provided in the book *Nutrition and Feeding of Organic Cattle* (Blair, 2011).

Ruminants differ from other farm animals such as pigs in having a more complex digestive system designed to allow the extraction of nutrients from bulky forages. This requires that they 'chew the cud', i.e. regurgitate ingested feed boluses from the stomach area back into the mouth

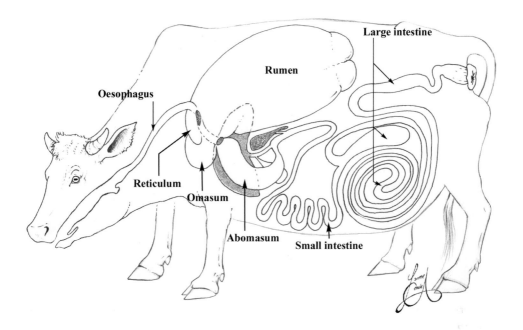

FIGURE 5.1 Digestive system of cattle.

for further chewing and grinding. Chewing the cud reduces feed particle size, and mixes saliva into the feed to assist in swallowing. In comparison with pigs they have no upper canine teeth or incisors, and have long, thick and rough tongues that are designed to optimize the prehension of forage. In addition, these animals possess a stomach compartment called a rumen, a fermentation organ populated by microorganisms that attack and break down the relatively indigestible feed components such as forage and roughages.

Digestion is the preparation of ingested material (ingesta) for absorption, i.e. reduction of feed particles in size and solubility by mechanical and chemical means. The final breakdown is achieved by enzymes secreted in the digestive juices and by gut microorganisms.

As in other farm animals, the alimentary system of cattle is composed of a mouth, tongue, teeth, oesophagus, stomach area, a small and large intestine, ancillary organs and a rectum (Figure 5.1). However, the stomach section is more complicated, comprising a reticulum, rumen, omasum and abomasum (true stomach). This modification of the digestive system, together with the secretion of large amounts of saliva, is an adaptation to a diet high in fibrous feedstuffs. The modification allows ruminant animals to pre-digest the feed, with around 60–75% of ingested material being broken down by chewing and microbial action before the ingesta enter the true stomach.

The rumen is a large fermentation chamber (in adult cattle about 125 litres) in which a large population of microorganisms (mainly bacteria and protozoa) attacks and breaks down the relatively indigestible feed particles by secreting enzymes capable of breaking down cellulose. In addition, these organisms synthesize nutrients such as B-vitamins and essential amino acids, which become available to the animal when the microorganisms die and are digested.

The by-products of microbial fermentation are volatile fatty acids (VFAs), mainly acetic, propionic and butyric acids, which are then absorbed to provide most of the animal's energy needs. The proportion of VFAs varies with diet, although the major product is always acetic acid. A high proportion of this fatty acid is important for milk fat synthesis. With a diet high in fibre, the molar ratio of acetic to propionic to butyric acids is about 70:20:10.

The rumen is not functional at birth and only becomes functional once the calf begins to eat fibrous feed. In cattle it is fully functional once the calf is about 3 months of age.

After fermentation, the ingesta pass to the abomasum, which corresponds to the stomach of non-ruminant animals such as pigs. This organ secretes the gastric juices, which contain several enzymes, principally pepsin, that act to break down protein to smaller units (peptides). A small amount of lipase present in gastric juice starts the digestion of fat in the abomasum. In nursing calves, the gastric juice also includes the enzyme rennin, which breaks down the protein in milk.

Stomach of the Newborn Calf

Calves (also lambs and kids) at birth are not functional ruminants. At this stage the rumen is very small and undeveloped. As a result, digestion is more like that of the pig.

Newborn ruminants for a short time after birth (up to 36 hours) have the ability to absorb large molecules via an oesophageal groove. This is important in that it allows newborns to receive immunoglobulin from colostrum (first milk from a nursing mother), which provides some immunity against diseases in the environment until active immunity is functional.

During the suckling process, impulses from the brain send messages to the oesophageal groove, causing the sides of the groove to curve upward and form a tube. It allows a direct flow of milk into the abomasum and secretion of the enzyme rennin from the wall of the abomasum, causing the milk to coagulate or curdle. This slows the passage of milk through the abomasum, allowing ample time for the milk to be digested. As the young animal gets older and starts to take in solid feed, the rumen begins to develop. This development is aided by the production of VFAs. By the end of the fourth week, for instance, the dairy calf should be able to utilize some grain and high quality hay.

Small Intestine

The small intestine is the location where final digestion of the ingesta occurs and absorption takes place. In form and function it is similar to that of the pig.

A main difference between the digestive process in ruminant animals and non-ruminant animals (such as pigs and chickens) is that much of the dietary carbohydrate in ruminants is broken down to VFAs rather than glucose.

As a result of these digestive and fermentation activities, the ingested carbohydrates, protein and fats are broken down into small molecules. Absorption of the final products of digestion takes place in the small intestine, so that digestion and absorption are complete by the time the ingesta have reached the terminal end of the small intestine ileum. This area is therefore of interest to researchers studying nutrient bioavailability (relative absorption of a nutrient from the diet) since a comparison of dietary and ileal concentrations of a nutrient provides information on its removal from the gut during digestion and absorption. Minerals and vitamins are not changed by enzymatic action.

They dissolve in various digestive fluids and water, and are then absorbed. Once the nutrients enter the bloodstream or lymph, they are transported to various parts of the body for vital body functions. Nutrients are used to maintain essential functions such as breathing, circulation of blood and muscle movement, replacement of worn-out cells (maintenance), growth, reproduction and secretion of milk (production).

The remaining ingesta, consisting of undigested feed components, intestinal fluids and cellular material from the abraded wall of the intestine, then pass to the next section of the intestine, the large intestine.

Here the intestinal contents move slowly and no enzymes are added. Some microbial breakdown of fibre and undigested material may occur, but absorption is limited. Remaining nutrients, dissolved in water, are absorbed in the lower part of the colon. The large intestine absorbs much of the water from the intestinal contents into the body leaving the undigested material, which is formed into the faeces and later expelled through the anus.

The entire process of digestion takes about 24–36 hours.

DIGESTION OF CARBOHYDRATES

Plant tissues contain about 75% complex carbohydrates (fibre), which provide the main source of energy for both the rumen microbes and the ruminant animal. About 30 to 50% of these carbohydrates is digested in the rumen by the microbial population. Some 60% or more of the starch is degraded, depending on the amount fed and how fast the ingesta move through the rumen. Most sugars are completely digested within the rumen.

During microbial digestion an appreciable amount of gas (mainly carbon dioxide and methane) is produced, representing about 6 to 7% of the feed energy of the ruminant. Under normal conditions, distension from gas formation in the rumen causes the cow to belch and eliminate the gas. Bloat can occur if the gas is not released.

As outlined above, the main end products of carbohydrate digestion are VFAs and when large amounts of forage are fed the formation of acetic acid is predominant. When grain feeding is increased or when finely ground forages are fed, the pattern of VFA production changes, with the proportion of acetic acid decreased and the proportion of propionic acid increased.

One of the by-products of fermentation of carbohydrates to VFAs is hydrogen, which is converted to methane gas before being released from the rumen. It has been shown that as the pattern of ruminal fermentation alters from acetate to mainly propionate, both hydrogen and methane production are reduced. This relationship between methane production and the ratio of the various VFAs has been well documented. It explains why the feeding of fibrous diets results in more methane than less fibrous diets. This aspect is becoming of increasing importance in relation to greenhouse gas production from agriculture. Methane is considered to have 21 times the global warming potential of carbon dioxide. As a result of these findings some researchers claim that organic milk production inherently increases methane emission because of the feeding system.

Carbohydrates such as sugars and starches that escape digestion in the rumen are digested in the abomasum, and the end products absorbed in the small intestine.

Although fibre is the most indigestible portion of the diet, it is necessary for the correct functioning of the ruminant gut. The amount recommended in the diet depends on the age and stage of production of the animal in question. For instance, dairy cows producing large amounts of milk should be fed diets with less fibre, while those producing less milk or are growing, or are dry, should be fed diets with more fibre from forage sources.

Energy is obtained when the feed is digested in the gut. The energy is then either released as heat or is trapped chemically and absorbed into the body for metabolic purposes such as maintenance, growth or production of milk and meat. It can be derived from protein, fat or carbohydrate in the diet. In general, forage products and cereal grains provide most of the energy in the diet. The provision of energy accounts for the greatest percentage of feed costs. Energy in excess of requirement is converted to fat and stored in the body.

The total energy (gross energy; GE) of a feedstuff can be measured in a laboratory by burning it under controlled conditions and measuring the energy produced in the form of heat. Digestion is never complete under practical situations; therefore measurement of GE does not provide accurate information on the amount of energy available to the animal. A more precise measurement of energy is digestible energy (DE), which takes into account the energy lost during incomplete digestion (i.e. GE minus energy in the faeces).

More accurate measures of useful energy contained in feedstuffs are metabolizable energy (ME; which takes into account energy lost in the urine) and net energy (NE; which in addition takes into account the energy lost as heat produced during digestion). Consequently, ME began to be used in ruminant nutrition and is still used in some feeding systems. Later the more meaningful NE came into use. NE is defined as ME minus the heat increment, which is the heat produced (and thus energy lost) during digestion of feed, metabolism of nutrients

and excretion of waste. The energy left after these losses have been deducted is the energy (NE) actually used for maintenance and for production (growth, gestation, lactation). Thus the NE system is the most exact system that describes the energy that is available to the ruminant animal.

NE is therefore used as the most accurate way to quantify the energy content of feeds. For ruminant feeding it has been refined into NEm, NEg and NEl, these being defined as follows:

Net Energy for Maintenance (NEm) is the amount of energy in a feed required to keep an animal in energy equilibrium, neither gaining weight nor losing weight.

Net Energy for Gain or Growth (NEg) is the amount of energy in a feed required for body weight gain once maintenance is achieved.

Net Energy for Lactation (NEl) is the amount of the energy in a feed required for maintenance plus milk production during lactation.

Digestion of Proteins

Dietary protein, like dietary carbohydrate, is fermented by rumen microbes. The main product is ammonia, organic (carbon-containing) acids, amino acids, and other products. Approximately 40 to 75% of the protein in feed is broken down in the rumen. Many rumen microorganisms require ammonia for growth and the synthesis of microbial protein.

Ammonia is most efficiently incorporated into bacterial protein when the diet is rich in soluble carbohydrates, particularly starch. Ammonia in excess of that used by the microbes is absorbed and mainly excreted in the urine. Some urea is returned to the rumen via the saliva.

Feed protein that escapes breakdown in the rumen (sometimes called by-pass protein) and microbial protein pass from the rumen to the abomasum, where they are digested and absorbed into the bloodstream through the wall of the small intestine. The fact that the protein passing into the abomasum is from two sources – microbial protein and by-pass protein (protein not broken down in the rumen) – makes measurements of protein digestibility difficult in ruminant animals.

Evaluating feedstuffs in terms of crude protein does not provide accurate information on how well the dietary protein is digested and provides an optimal supply of amino acids for maintenance, growth and production of milk, meat and wool. Feedstuffs vary greatly in protein digestibility. For example, the digestibility of the protein in common cereal grains and most protein supplements is around 75–85%, while in alfalfa hay it is around 70% and in grass hay is around 35–50%. Thus, even though total protein intake may appear to be adequate, the animal may be deficient in this nutrient. Therefore, more accurate measures of protein quality have been introduced for ruminant feeding. One of these is **Metabolizable Protein (MP)**, now used in the rationing systems for ruminants.

MP is defined as the true protein absorbed in the form of amino acids in the small intestine. It is a calculated value based on a database of research findings related to the extent of undegradability of various protein feedstuffs in the rumen (also called by-pass protein) and the amount of microbial protein produced in the rumen. The term undegradability refers to the extent of resistance to breakdown in the rumen, with the result that the protein can pass to the small intestine intact.

Some ruminant animals such as growing heifers, dry cows, and cows in mid to late lactation may meet their MP needs solely from microbial protein produced in the rumen. However, high-yielding dairy cows have amino acid requirements that cannot be met from microbial protein alone and require a supplement of protein in the diet. In this situation the diet should include proteins of low degradability in the rumen so that they escape breakdown until they reach the small intestine. This escape or by-pass protein is now termed **Rumen Undegraded Protein (RUP)**.

Digestion of Fats

Most of the digestion and absorption of fats occurs in the small intestine. Rumen microorganisms convert unsaturated fatty acids to saturated acids by the addition of hydrogen molecules. Thus, more saturated fat is absorbed by ruminant animals than by non-ruminant animals such as pigs.

VITAMIN SYNTHESIS

For the first few weeks of life, calves, lambs and kids are essentially non-ruminant animals and have dietary requirements similar to those of pigs and poultry. Initially, therefore, they must obtain all the required nutrients from milk or milk replacer. They require high quality, easily digested feeds to supply needed energy, essential amino acids, essential minerals and vitamins. After about 5–6 weeks of age, forage and grain consumption increases and microorganisms in the rumen become increasingly active in synthesizing essential amino acids and B vitamins and in digesting fibre. When the rumen is fully functional the ruminal microorganisms synthesize all the B vitamins and vitamin K required by cattle, at least for growth and maintenance. Therefore, ruminating cattle should not require supplementation with B vitamins or vitamin K.

COMPONENTS OF ORGANIC DIETS FOR RUMINANT ANIMALS

Forage

Grazing. One of the leading countries in organic production is Germany. As reported by Haas et al. (2007), organic dairy farms in that country are not yet at the stage of feeding the cows on all-forage diets. This may be related to farm size in Germany. Milk yield was found to be almost 7,000kg per ha on the organic farms. The researchers calculated that 0.96ha per cow was needed to produce the feed requirement for that level of production, of which 0.85ha was farmland and the production area for purchased feed was 0.11ha. Their data showed that on an energy basis (MJ NEL), 74% of the annual average milk yield of 6,737kg per cow was derived from forage, 23% from purchased concentrates, and 3% from commercial processing by-products such as spent grains from the brewing industry.

A 2000–01 Economic Farm Survey (Verkerk and Tervit, 2003) found that the dry matter intake of the average New Zealand dairy cow on conventional farms comprised a higher content of forage, being made up of 88.5% grazed pasture, 5.5% pasture silage, 3.0% maize silage, 2.0% purchased grazing and only 1.0% supplement.

These data confirm that, while forage is the main feed of organic cattle, supplementary feeding with grains and other feedstuffs is usually necessary in many countries, especially for dairy cows. Weller (2002) carried out a comparison between two systems of organic dairy farming, one with a high stocking density using purchased concentrates, and the other a self-sufficient system.

He found that the self-sufficient system had more problems in balancing the dietary energy, resulting in lower milk production, more post-calving health problems and a reduction in reproductive performance.

In countries with temperate climates grazed forages are utilized in late spring, summer and early autumn, while some regions, such as Australia–New Zealand and South America, may support cattle production on year round grazing of forages. At other times of the year conserved forages have to be fed. In some regions the forage may be deficient in certain trace elements, requiring that the deficiency be remedied by providing the necessary nutrients in the form of a supplement, feed blocks or mineral licks. For instance, in parts of North America the soils and pastures are low in iodine; in this case the use of iodized salt is recommended.

Pasture is generally based on grasses (e.g. perennial ryegrass, *Lolium perenne*) with a legume such as white clover (*Trifolium repens*) included in the mix to fix atmospheric nitrogen and improve the nutritive quality of the forage. When young and lush such forage is a feed of high nutritive value and may provide most of the requirements of a good dairy ration. Higher milk yields usually require supplementary feeding, especially when the pasture is of lower quality.

It is necessary for the forage, whether grazed or conserved, to be of high quality in order for acceptable milk yields to be achieved. In addition to the mix of plants forming the sward, forage quality is also determined by its stage of development and by the soil and climatic conditions. Utilizing or harvesting the stand of forage at the correct stage of growth is an important first step in attaining high quality.

As Kersbergen (2010a) has pointed out, nutritionists characterize quality forages as feeds that provide high levels of digestible nutrients and have the potential for high intakes by cattle, while maintaining ruminal health. Intake potential is a good barometer of quality in forages, since maximizing forage intake will result in healthy cows with good milk production. As plants become larger, cell contents (which are 100% digestible) decline with an associated increase in cell wall content. As the cell walls become a larger proportion of the forage with increasing maturity, the percentage of lignin (100% indigestible) increases and the ability of the plant material to be digested by microorganisms in the rumen declines. As the plant continues to mature, the concentration of protein in the plant material declines as well. To add to the situation, as forages become more fibrous and less digestible, they also decrease the ability of animals to consume large amounts of feed, further reducing nutrients that are available for milk production. Maximizing intake from quality forages should be a priority for all dairy producers to maintain good production levels and body condition.

Research in Ireland (Dillon, 2009) shows that grass production is maximised by grazing to 3.5–4cm residual height. With good quality grass this should yield 1,250kg/DM per ha. By keeping the pasture in a growing state, a higher quality of grass will be produced in a green leafy base. Pre-grazing height should be 8–9cm (three leaves), if this is grazed down to 3.5–4cm then growth will be 16 tonnes/ha. Allowing the grass to go to seed should be avoided. An electric fence can be used to allocate grass on a 12-hour basis when all the available grassland is being used. The research also showed that when cows are restricted to two 3-hour periods of grazing, 97% of that time is spent grazing. And when cows are given 24-hour access to grazing only 41% of the time is spent grazing. The target pre-grazing yield should be between 1,200 and 1,500kg DM/ha.

Many organic producers hold the view that acceptable milk production can be obtained with forage alone. It is useful therefore to review the results of experiments in which this issue was put to the test.

Data provided by Stockdale (1999) from Australia are relevant in this connection. The study in question involved three short-term experiments with Friesian cows provided with pasture only or a supplement of 5kg (DM) per cow per day of pelleted cereal grain (75% barley and 25% wheat), pelleted mixed grains (50% lupin seed, 25% barley and 25% wheat) or hay. The hay used in experiment 1 was made from lucerne, while that used in experiments 2 and 3 was from irrigated annual and perennial pastures, respectively. Cows strip-grazed irrigated pasture at a herbage allowance of about 30kg DM/cow/day in each experiment. Prior to each experiment the average milk yield of the cows was 30, 25.6 and 16.9kg per day, respectively. Days in lactation were 105, 114 and 222, and the cows were 6, 6 and 7 years of age, respectively. Results showed that supplementation was beneficial, all supplements resulting in a significant increase in milk production. The lupin seed + cereal grain supplement gave the highest response and the hay the lowest response, both in terms of yield and the marginal return to additional total dry matter consumed (1.4, 1.7 and 0.9kg of extra milk for each additional kg dry matter from each supplement, respectively).

Yield of milk solids responded similarly to milk yield.

In some regions a limiting factor may be that the forage is deficient in certain trace elements, requiring that the necessary nutrients be supplied in the form of supplementary feed, feed blocks or mineral licks. Local advice should be obtained on this issue.

The pasture is generally based on grasses (e.g. perennial ryegrass, *Lolium perenne*) with a legume such as white clover (*Trifolium repens*) included in the mix to fix atmospheric nitrogen and improve the nutritive quality of the forage. When young and lush, such forage is a feed of high nutritive value and may provide most of the requirements of a good dairy ration.

Forage species, agronomic conditions, fertilizer practices, maturity at harvest and storage procedures are among the factors that determine the quality of the forage when fed to the animal. The most important grasses worldwide are orchard grass, ryegrass, fescue and timothy.

Legumes are also used as forage crops. The major species used worldwide are lucerne, clovers and birdsfoot trefoil. Under ideal conditions clovers can add up to 240kg nitrogen/ha/year to the soil, which can be utilised by other forage species. Legumes have a lower content of fibre and a higher content of crude protein than grasses. Thus, legumes are generally higher in feed value than grasses and when included in the forage increase the forage quality of pastures, hays or silages. Also, intake is generally higher when cattle are pastured on grass/legume mixtures or fed grass/legume hays than when they are fed grass alone. As a result, animal productivity often improves when a clover is included in pastures, even though total forage yield may not increase.

Clovers are implicated in some health problems for livestock, e.g. bloat, but generally problems only occur when the proportion of clover in the stand is more than 50%. A few clovers synthesize oestrogen-like compounds called phytoestrogens that can cause reproductive problems in livestock, more especially in sheep.

Clovers also help to extend the grazing season by allowing grazing at a time when other forages are not as active. In addition to the mix of plants forming the sward, forage quality is also determined by its stage of development and by the soil and climatic conditions. The grazing management system should, therefore, ensure that the forage is grazed at the stage when its nutritive value is highest, otherwise it may only supply the maintenance needs of the animal.

Lucerne

Lucerne (*Medicago sativa*, also known as alfalfa) is the most important forage legume worldwide. It can be grown over a wide range of soil and climatic conditions and has the highest yield and feeding value of all perennial forage legumes. This crop can be used for pasture or fed in cut or conserved forms such as hay, silage, green chop, and processed products such as meal, pellets and cubes.

Grazing the early spring growth provides quality feed and delays the first hay cut until weather conditions are more favourable for harvesting. Grazing during midsummer can provide forage when cool season grasses are often less productive. In grazing trials and demonstrations, the forage quality of alfalfa pasture has been found to be excellent, resulting in total season average daily gains of more than 1kg/day. In addition, milk yield from dairy cows is greater when these animals graze lucerne rather than grass.

Grazing can extend the useful life of a stand by 1 year or more for mature lucerne hay fields where some of the stand has been lost or has become weedy. Grazing may also rejuvenate some stands by reducing grass and weed competition. Research has shown that lucerne stands with fewer than 30 plants per square metre may not produce maximum yields of hay.

Alternative Temperate Forages

Chicory (*Chicorium intybus*), legumes containing condensed tannins (*Lotus corniculatus*) and sulla (*Hedysarum coronarium*) are also useful forage plants. An interesting finding with these species was that chicory and sulla promoted faster growth rates in young sheep and deer infested with internal parasites. Grazing on chicory forage has been shown to increase the reproductive rate in sheep, increase milk production in both ewes and dairy cows and reduce methane production, effects that were mainly due to its content of condensed tannins (CT). Risk of rumen frothy bloat in cattle grazing legumes was shown to be reduced when the forage contained at least 5g CT/kg (DM basis).

Conserved Forages

Green chop is very similar to grazed forage except that a machine is used to harvest the crop. Harvesting and storage losses are generally very low. However, equipment and energy usage and labour costs are high. Harvesting and storage losses are greatest with hay and silage, but if proper practices are followed these losses can be minimized.

All the species mentioned above are suitable for organic feeding. Grasses such as Bermuda grass (*Cynodon dactylon*) that are not mentioned specifically in the lists of approved organic feedstuffs (Chapter 2) are probably acceptable for organic diets, but this should be checked with the local certifying agency.

As stated in Chapter 2 in the section relating to forages and roughages, only the following substances are included in this category: lucerne, lucerne meal, clover, clover meal, grass (obtained from forage plants), grass meal, hay, silage, straw of cereals and root vegetables for foraging. They can be conserved by haymaking, silage, etc. As worded, this section applies only to harvested forages and does not apply to grazed forage.

Hay

Haymaking is a traditional method of conserving green crops and is popular with organic producers in Western Europe and other regions. It is made mainly by the sun drying of grass and other forage crops. After the crop has been cut, its treatment in the field is intended to minimize losses of valuable nutrients caused by the action of plant respiration, microorganisms, oxidation, leaching and by mechanical damage. The aim in haymaking is to reduce the moisture content of the cut crop to a level low enough to inhibit the action of plant and microbial enzymes and allow the hay to be stored satisfactorily for later feeding.

During the early phase of drying, enzymes break down or reduce simple sugars and organic acids to carbon dioxide, resulting in an overall loss of dry matter and digestible components. Hence there is a need for drying to be as rapid as possible. Losses of forage dry matter in the field can range from less than 5% to more than 50%, depending on weather conditions and how long it takes the plants to dry. In warm, dry, windy weather the wet herbage, if properly handled and mechanically agitated, will dry rapidly and losses arising from plant enzyme activity will be low. Some producers use field machinery and barn drying equipment to speed up the process.

Even under good conditions the overall loss of dry matter may be about 20%. Rainfall can leach plant protein, phosphorus, potassium, carotene and digestible energy components during hay cutting and drying.

The moisture content of a green crop may range from about 650 to 850g/kg, falling as the plant matures. For satisfactory storage the moisture content must be reduced to less than 200g/kg. It is not advisable to wait until the crop is mature and drier before cutting. The more mature the crop, the lower is its nutritional value. Moisture content can be measured by taking a sample from the windrow and drying it using a microwave or convection oven. Wet and dry weights can be measured with a scale.

Hay may become mouldy if not dried sufficiently. Mouldy hay is unpalatable to livestock and may be harmful to farm animals and humans because of the presence of mycotoxins. Such hay may also contain actinomycetes, which are responsible for the allergic disease affecting humans known as 'farmer's lung'.

A traditional method of haymaking, which is still practised in some parts of the world such as Switzerland, Italy, Germany and Scandinavia, is to make hay on racks, frames or tripods. Hays made by tripoding are generally of higher quality than hays made by ground drying, especially in wetter climates. Rack and tripod drying also allow a younger forage to be cut.

Various types of machinery can be used to crimp and crush the crop during the haymaking process and speed up the rate of drying. Leaves of forage plants dry more quickly than stems due to the higher water content of stems. Orchard grass, tall fescue and timothy dry faster and more uniformly than legumes, clovers and ryegrasses. Legume leaf surfaces are more waxy than most grasses, resulting in a slower rate of drying.

During the drying process the leaves lose moisture more rapidly than the stems, becoming brittle and readily shattered by handling. If the herbage is bruised or flattened, the drying rates of stems and leaves are more similar. Excessive mechanical handling is liable to cause a loss of leafy material, and since the leaves at the hay stage are higher in digestible nutrients than the stems, the hay produced may be of low nutritive value. Loss of leaves during hay-making is particularly likely to occur with legumes such as lucerne. Machines are now available that reduce the losses caused by leaf shattering.

Overall losses during haymaking can be appreciable under poor weather conditions. In a study on 6 commercial farms carried out over a 3-year period in north-east England by the Agricultural Development and Advisory Service (ADAS, 2005), losses of nutrients were measured between harvesting and feeding. Total dry matter losses averaged 19.3%, made up of 13.7% field loss and 5.6% loss in the bale. The losses of digestible organic matter and digestible crude protein were both about 27%.

Prolonged drying of hay increases the loss of some vitamins and pigments. Carotene, a precursor to vitamin A, is unstable in the presence of sunlight and losses can be high if the hay is not dried quickly. On the other hand, sunlight increases the vitamin D content of hay due to irradiation of the precursor (ergosterol) present in the green plants.

Artificial drying (known commercially as dehydration) is a very efficient but expensive method of conserving forage crops. It tends to be a commercial process rather than one found on organic farms. In northern Europe, grass and grass/clover mixtures are the most common crops dried by this method, whereas in North America lucerne is the primary crop that is dehydrated.

The nutritive value of hay is determined by the stage of growth when it is cut and by the plant species of the parent crop. Yields are higher with late cuts but the nutritional value and voluntary intake by cattle are lower. Thus, hay made from early cuts is invariably of higher quality than hays made from mature crops.

As stated above, hays made from legumes are generally higher in protein and minerals than grass hay. Lucerne is grown as a hay crop in many countries, its value derived mainly from its relatively high content of crude protein, which may be as high as 200g/kg dry matter if it is made from a crop cut in the early bloom stage.

The nutritive value of hay is also affected by field losses of nutrients and by changes taking place during storage. Even under good conditions the overall loss of dry matter may be about 20%. Artificially dried forages are higher in nutritive value than hays. However, they are expensive to produce and tend to be used mainly in poultry and pig diets as a source of minerals and vitamins.

One point to note is that most weeds are not palatable and in pasture will be avoided by livestock if adequate forage is available. However, most livestock cannot differentiate weeds from beneficial long-stemmed forage in hay, resulting in accidental ingestion and possibly a loss in productivity or death. Thus, an effective weed control programme is required.

Hay preservatives may be used to allow hay to be stored at moisture levels that would otherwise result in severe deterioration and moulding. These chemical preservatives include propionic acid, lactic acid-forming bacteria and other biological products. They may be acceptable in organic hay production but this should be checked with the local certifying agency.

Straws consist of stems and leaves of plants after the removal of the ripe seeds by threshing, and are produced from most cereal crops and from some legumes. Chaff consists of the husk or glumes of the seed, which are separated from the grain during threshing. Modern combine harvesters put out straw and chaff together, but older methods of threshing (e.g. hand threshing) yield the two by-products separately. All the straws and related by-products are extremely fibrous. Most have a high content of lignin, and all are of low nutritional value. In rice-growing regions rice straw is used as ruminant feed. It is similar to barley straw in nutritional value. In contrast to other straws, the stems are more digestible than the leaves. A by-product similar in composition to straw is sugarcane bagasse, used for ruminant feeding in tropical countries. Of the straws likely to be available on organic farms, oat straw is preferred for cattle feeding. Apart from the low digestibility of these cereal straws, a major disadvantage is the low intake obtained when they are fed to ruminant animals. Whereas a cow will consume up to 10kg of medium quality hay, it may eat only about 5kg of straw. Improvements in both digestibility and intake of straw can be obtained by the addition of protein to the feed mixture.

Cereal hay is suitable as the forage component of rations for all classes of beef cattle, sheep and dairy cattle and should be equal in value to good quality bromegrass hay.

Fresh cereal straw is a good alternative in wintering feed for cows and sheep if properly supplemented with an energy source such as grain and added minerals and vitamins. Straw can be used in combination with other feeds as the sole roughage for beef cows. However, its use with dairy cows should be limited to 4–5kg/day for satisfactory milk production.

Silage

Ensilage is the name given to the process of conserving a crop of high moisture content by controlled fermentation. The product is known as silage. Almost any crop can be preserved as silage, but the most common are grasses, legumes and whole cereals, especially wheat and maize.

There are two major objectives in making hay or grass silage. The first is to remove excess herbage from pasture following its rapid growth in the spring, allowing the land to be grazed subsequently without wastage of surplus forage. The second objective is to conserve the material so that it provides a nutritious feed for cattle when grazing is limited or unavailable. Silage production is an alternative to haymaking. Often it is difficult to make hay satisfactorily because of climatic conditions. For instance, in order to produce good grass hay it is necessary to reduce the moisture content to less than 16% otherwise moulds can develop during storage.

To ensure a stable fermentation, the ensiled material is stored in a silo or other container and sealed to maintain anaerobic conditions. The three most important requirements for good silage production are: (i) rapid removal of air; (ii) rapid production of lactic acid, which results in a rapid drop in pH; and (iii) continued exclusion of air from the silage mass during storage and feeding. In practice this is achieved by chopping the crop during harvesting, rapid filling of the silo and adequate consolidation and sealing. After chopping, plant respiration continues for several hours and plant enzymes such as proteases are active until all the oxygen is used up. These enzymes break down the protein in the forage. Rapid removal of air is also important because it prevents the growth of undesirable aerobic bacteria, yeasts and moulds that compete with beneficial bacteria for substrate.

Fermentation begins once anaerobic conditions are achieved. Aerobic fungi and bacteria are the dominant microorganisms on fresh herbage, but as anaerobic conditions develop in the silo they are replaced by bacteria that are able to grow in the absence of oxygen. During the ensilage process these lactic acid bacteria continue to increase, converting the water soluble carbohydrates in the crop to organic acids (mainly lactic acid). A rapid reduction in silage pH helps to limit the breakdown of protein in the plant material by inactivating enzymes in the forage that break down proteins. In addition, a rapid decrease in pH inhibits the growth of undesirable anaerobic microorganisms such as enterobacteria and Clostridia. Eventually, continued production of lactic acid and a decrease in pH inhibits growth of all bacteria.

In general, good silage remains stable, without a change in composition or temperature once oxygen is eliminated and the silage has achieved a low pH.

Achieving the critical pH is more difficult with legume crops, which require a higher level of acid production to lower the pH than maize silage.

The dry matter content of the forage can also have a major effect on the ensiling process. Wet crops are very difficult to ensile satisfactorily. Drier silages do not pack well, making it difficult to exclude all the air from the forage mass. Also, as the dry matter content increases, growth of lactic acid bacteria is curtailed and the rate and extent of fermentation is reduced. On the other hand, wilting forage above 30–35% dry matter prior to ensiling can reduce the incidence of undesirable organisms such as Clostridia.

Various silage additives have been used to improve the nutrient and energy recovery in silage, often with subsequent improvements in animal productivity. These include live organisms such as lactobacilli, enzymes and propionic acid. Molasses, which is a by-product of the sugar beet and sugarcane industries, was one of the earliest silage additives. Producers need to check with the local certifying agency to determine whether these additives are acceptable for organic production in their region.

The nutritional value of silage depends upon the species and stage of growth of the harvested crop, and on changes that occurred during the harvesting and ensiling period. For instance, maize is best cut for silage when at the dent stage of maturity. Thus silage can be quite variable in quality. Information on the moisture, digestible organic matter and ammonia nitrogen contents is important prior to feeding since these have been shown to be the major determinants of silage intake. Periodic sampling is therefore advised at co-operative, government or commercial laboratories.

The highly degradable nature of the nitrogen in most silages points to the need for adequate supplementation of silage-based diets with a readily available supply of carbohydrate, so that the rumen microbes can assimilate the rapid influx of ammonia following an intake of silage. This maximizes the synthesis of microbial protein and minimizes the loss of both nitrogen and energy. Silage-based diets must, therefore, contain a supplemental source of energy to maximize the utilization of the nitrogen of the diet. A similar benefit has been obtained by supplementing silage diets with soya bean meal, presumably by making amino nitrogen available to rumen microorganisms, which would otherwise be dependent upon ammonia as their main source of nitrogen.

An important aspect of ruminant feeding is that, since the various feed components are digested by different classes of microbes, allowance has to be made for new populations of microflora to establish in the gut whenever changes are made to the diet. A change in diet should, therefore, be made gradually otherwise a digestive disorder may occur. It may take up to 6 weeks for the rumen microorganisms to adapt to a change in diet.

FEED ANALYSIS

Since ruminant animals are fed forage-based diets and since forages vary widely in nutrient composition, it is desirable that the nutrient composition of a forage (either fresh or conserved) be known, to allow satisfactory rations to be formulated. This information can be obtained by a laboratory analysis.

The information contained in a lab report is mainly as follows:

1. Moisture (water): This can be regarded as a component that dilutes the content of nutrients and its measurement provides more accurate information on nutrient contents.

2. Dry matter (DM): This is the amount of dry material present after the moisture (water) content has been deducted.

3. Ash: This provides information on mineral content after the sample has been incinerated. Further analyses on the ash can provide exact information on specific minerals present.

4. Organic matter: This is the amount of protein and carbohydrate material present after ash has been deducted from DM.

5. Crude protein: This is determined as nitrogen content (N) × 6.25. It is a measure of protein present, based on the assumption that the average nitrogen content is 16g N/100g protein. Some of the nitrogen in most feeds is present as non-protein nitrogen, therefore the value calculated by multiplying nitrogen content by 6.25 is referred to as crude rather than true protein.

6. Non-nitrogenous material.

6.1. Fibre: Initially fibre was measured as a single fraction, crude fibre (CF). Now the detergent analysis system is used to separate the two main types of fibre in forage since some of the structural carbohydrates making up the fibre are more difficult to digest than others.

6.1.1. The fraction of a forage or feed sample that is insoluble in a neutral detergent solution is termed neutral detergent fibre (NDF). This fraction is composed of the primary components of the plant cell wall, namely hemicellulose, cellulose, and lignin.

6.1.2. Another measure of fibre is acid detergent fibre (ADF), part of the NDF. It is composed of the cell wall components that are more difficult to digest, namely cellulose, lignin and other related compounds. As forages mature, the cell walls become more lignified and less digestible. Acid detergent fibre content is therefore a good indicator of lignification.

7. Fat: Measured as crude fat (sometimes called oil or ether extract since ether is used in the extraction process). True fats provide about 2.25 times as much energy per unit of weight as carbohydrates. A point to note is that the method extracts other compounds from green forages and shrubs that do not provide any energy to the animal. More detailed analyses can be done in the lab to measure individual fatty acids.

8. Minerals: The content of major minerals such as calcium and phosphorus can be measured as well as the content of trace minerals such as copper.

Some commercial laboratories are able to provide very detailed information on forages, for instance contents of dry matter, crude protein, moisture, soluble protein, ADF-CP (a measure of by-pass protein), degradable protein, ADF, NDF, total digestible nutrients (TDN, used in some rationing systems), NEl, NEg, NEm, relative feed value (RFV, used in some rationing systems), non-fibre carbohydrates, starch, lignin, calcium, phosphorus and other minerals.

Vitamins are not measured directly in current analysis systems but can be measured by appropriate methods.

Techniques such as near infrared

reflectance spectroscopy (NIRS) are now being used increasingly to replace the older and slower wet chemistry methods of feed analysis.

USING THE VALUES OBTAINED FROM LABORATORY ANALYSIS

Dry matter (DM) is the moisture-free content of the sample. Moisture dilutes the concentration of nutrients, therefore ruminant diets are formulated on a dry matter basis to ensure the correct balance and intake of nutrients.

DDM. Some feeding systems use digestible dry matter (DDM). This can be calculated from ADF, using a database of established relationships. It is a measure of the proportion of a forage that is digestible.

Energy

NDF. This fraction is a measure of the structural components of the forage, namely cell wall material. NDF is, therefore, a good measure of plant maturity and feed quality. The concentration of NDF in feeds is negatively correlated with energy concentration. Since it provides bulk or fill to the diet it can be used to predict the voluntary intake of feed. In general, low NDF values are preferred because NDF increases as forages mature. As NDF content of a feed increases, dry matter intake decreases and chewing activity and rumination increase. NDF values are important because they allow the amount of forage the animal can consume to be calculated (using established equations). For legume forages, an NDF content less than 40% indicates good quality, while a content higher than 50% indicates lower quality. For grass forages, an NDF content

of less than 50% indicates high quality and a content of more than 60% indicates lower quality.

ADF. This fibre fraction is part of the NDF and is composed of the cell wall components (cellulose, lignin, silica and insoluble forms of nitrogen) that are more resistant to digestion. As ADF content increases, the digestibility of the forage decreases. Thus the ADF content can be used to estimate the energy value of feeds, using established equations. Like NDF, ADF is a good indicator of feed quality, higher values within a feed indicating lower quality feed. Generally a target value of less than 35% ADF in either legume or grass forages is recommended.

Tables of feedstuff composition increasingly quote NDF and ADF values rather than the older CF values, since NDF and ADF values are now used in practical feeding programmes. It is important to note, however, that CF is still the fibre component required by feed regulatory authorities to be stated on the feed tag of purchased feed, at least in North America.

Metabolizable energy (ME): As outlined above, this is measured as digestible energy (DE) minus the energy lost in the faeces). ME is used in some countries in ruminant feeding systems, the ME content of feedstuffs being calculated from the content of ADF using a database of established relationships.

Net energy (NE). As noted above, NE values are of most value in setting the energy level of ruminant diets. The values used with mature animals (NEm), growing animals (NEg) and lactating females (NEl) are not measured directly by analytical laboratories and instead are calculated from ADF values using a database of relationships established in research studies.

Total digestible nutrients (TDN): The sum of the digestible fibre, protein, fat, and carbohydrate components of a feedstuff or diet. TDN is directly related to digestible energy and is often estimated from ADF. Beef, sheep and goat producers in North America commonly use TDN in formulating feed rations.

Protein. As noted above, crude protein (CP) measures the nitrogen content of a feedstuff, including both true protein and non-protein nitrogen. For ruminant feeding it is useful to have information on the fraction that is degradable in the rumen (degradable intake protein, DIP) and the rumen-undegradable fraction (undegradable intake protein, UIP) since this fraction is utilized directly by the animal and absorbed as small proteins and amino acids. However, the rumen degradability of protein is not measured in most laboratories. Instead, diets are commonly formulated using analyzed CP values and average values for DIP and UIP published in feed composition tables.

Metabolizable protein (MP). MP is a measure of the amount of protein that is available to the animal, including microbial crude protein (BCP) synthesized by the rumen microorganisms and UIP. MP is defined as the true protein absorbed in the form of amino acids in the small intestine. It is a calculated value based on a database of research findings related to the extent of undegradability of various protein feedstuffs in the rumen (UIP, also called by-pass protein) and the amount of microbial protein produced in the rumen.

OTHER APPROVED FEED INGREDIENTS

Cattle, sheep and goats at certain stages require other feedstuffs in addition to forage. This is especially the case with young animals and high-producing dairy cows.

Based on the information in Table 2.1, which is drawn from both the northern and southern hemispheres, the following can be suggested as a potential list of the main feedstuffs available for the supplementary feeding of organic cattle, sheep and goats in many countries. Not all the feed ingredients listed as permitted ingredients are suitable for inclusion in ruminant diets, since the lists include those more suited for poultry and pig feeding. In addition, some of the ingredients are not usually available in sufficient quantity.

Due to a lack of data on feedstuffs that have been grown organically, the data are derived mainly from feedstuffs that have been grown conventionally. It is inferred that organic feedstuffs are similar in composition and nutritive value to conventional feedstuffs, except where a difference is stated. In time, a database of organic feedstuffs will be developed.

CEREALS, GRAINS, THEIR PRODUCTS AND BY-PRODUCTS

The Canadian Organic Standards are typical in allowing no more than 40% grain (Section 6.4.3 b for ruminants) in the diet and specifying that at least 60% of dry matter in daily rations must consist of hay, fresh/dried fodder or silage.

Organic producers are most interested in those grains that can be grown on-farm. Cereal grains are mainly energy sources, the main component of the dry matter being starch. Of the grains used as feed, oats have the lowest energy value and maize has the highest. The total content of protein is very variable, ranging usually from 80 to 120g/kg dry matter. All grains are deficient in calcium, containing less than 1g/kg dry matter. The phosphorus content is higher at 3–5g/kg dry matter and although part of this is present as phytate, ruminant animals produce a phytate enzyme in the rumen that breaks down phytate-bound phosphorus and makes it available to the animal. The cereal grains are deficient in vitamin D for ruminants and, with the exception of yellow maize, in provitamin A (β-carotene). Cereal grains are good sources of vitamin E for ruminants.

Cereals generally form a low proportion of the total diet of cattle, although they are the major component of supplementary rations. Calves often depend upon cereal grains for their main source of energy, and at certain stages of growth as much as 90% of their diet may consist of cereals and cereal by-products.

It is common for heat processing to be used commercially to improve the nutritive value of cereal grains. Steaming and flaking are known to increase the proportion of propionic acid in the volatile fatty acids produced during digestion. While about 75% of the starch of ground maize is digested in the rumen, this is increased to about 95% following steaming and flaking. Even greater effects have been recorded with sorghum (ground 42% versus steam processed 91%).

Oats

Oats have traditionally been used as a grain for feeding to cattle, but are no longer the predominant grain used. Higher energy cereals such as barley and maize are now more popular for the supplementary feeding of dairy cows and beef cattle. An advantage in using oats is that little or no processing is required. Oats are lower in energy and more bulky than other common feed grains because the hull remains on the grain after harvesting. Oats usually contain 110–140g/kg crude protein on a dry matter basis. The neutral detergent fibre content may exceed 300g/kg (dry matter basis). The acid

detergent fibre content is lower at around 100–150g/kg.

Barley

Barley is the grain most widely used in the supplementary feeding of ruminant stock in northern Europe, Canada and in the northern USA. It is widely used in feed mixtures for all types of dairy animals, including young calves and growing animals as well as lactating and non-lactating dairy cows. Barley contains (dry matter basis) about 646g/kg starch, compared with maize at 757, wheat at 700 and oats at 580. The protein content ranges from about 60 to 160g/kg dry matter, with an average value of about 120g/kg.

Barley starch is known to be digested more rapidly than maize starch, therefore care has to be taken to avoid a condition known as rumen acidosis (rapid fermentation of the starch to lactic acid) that causes a depression of digestion and of fibre and feed intake). Bloat can result. Therefore it is necessary to introduce high-concentrate feed supplements to ruminants gradually over a period. Coarse grinding is also recommended over fine grinding to reduce the speed of fermentation in the rumen.

Tempered rolled barley is often the preferred method of processing barley for dairy cows. This process involves adding water to bring the moisture content to 180–200g/kg, and allowing it to temper for 24 hours prior to rolling unless a wetting agent is used. The large number of small particles or 'fines' produced by dry rolling or grinding provide more surface area for starch digestion to occur, resulting in increased rate of starch degradation in the rumen. Fewer small particles are produced with rolled tempered barley compared to dry rolling, resulting in a reduced rate of fermentation. Rapid fermentation can lead to reduced pH and acidosis conditions in the rumen. Compared with dry rolled barley, tempering has been shown to improve milk yield by 5%, feed efficiency by 10%, apparent digestibility of dietary dry matter by 6%, NDF digestibility by 15%, ADF digestibility by 12%, CP digestibility by 10% and starch digestibility by 4%. Tempering is also recommended when barley is fed whole, since whole kernel digestibility is greater than with dry, raw grain. The explanation for the improved digestibility is that the rapid rate of passage in mixed diets with substantial amounts of forage allows little time for breakdown of the intact kernels.

Research has been conducted to test the effect of different grains (barley, maize, wheat, sorghum) on forage intake and digestion in beef steers fed grass hay during periods of feed shortage. No differences in hay intake or digestibility were noted when these grains were formulated into the diets based on their nutritional characteristics. Barley has also been shown to give improved production and prevent loss of condition in breeding herds on range.

Whole barley is not well utilized by beef cattle and a degree of processing is advised. Tempering and rolling generally results in a marked increase in digestibility. Owing to the rapidly fermentable nature of barley, the grain should only be coarsely cracked, not finely ground.

Young stock benefit also when barley is processed.

Maize

This cereal is also known as corn in the Americas, and can be grown in more countries than any other grain crop because of its versatility. It is the most important feed grain in several countries because of its palatability, high energy value and high yields of digestible nutrients per unit of land. As a consequence, it is used as a yardstick in comparing other feed grains for animal feeding. A number of different types of maize exist and the grain appears in a variety of colours, including yellow, white or red. Yellow maize is the only cereal grain to contain vitamin A, due to the presence of provitamins (mainly β-carotene). Yellow maize tends to colour the fat in beef, though the effect is less than that produced by grass.

Maize is an excellent energy source for cattle but it is low in protein, averaging about 85g protein per kg. It contains about 730g starch/kg dry matter, is very low in fibre and has a high energy value. The starch in maize is digested more slowly in the rumen than the starch in other grains, and at high levels of feeding a proportion of the starch passes into the small intestine, where it is digested and absorbed as glucose. This has advantages in treating conditions such as ketosis. When the starch is heated during processing it is readily fermented in the rumen.

The quality of maize is excellent when harvested and stored under appropriate conditions, including proper drying to 100–120g/kg moisture. Fungal toxins (zearalenone, aflatoxin and ochratoxin) can develop in grain that is harvested damp or allowed to become damp during storage. These toxins can cause adverse effects in livestock.

Maize is suitable for feeding to all classes of ruminants. It should be ground medium to medium-fine and should be mixed into feed immediately after grinding, since it is likely to become rancid during storage.

Maize By-products

Several by-products are obtained in the manufacture of starch and glucose from maize that are used in the feeding of farm animals and may be acceptable by organic certification agencies. One of the benefits of using maize by-products is that the grain used is of very high quality since the main product is intended for the human market. This helps to ensure the maize is free from mycotoxin contamination and insect and rodent infestation.

Maize gluten feed is a maize by-product that is available in some regions as a wet or dry product. The dry product is traded internationally. This feed has a variable protein content, normally in the range 200–250g/kg dry matter, of which about 60% is degraded in the rumen. It has a crude fibre content of about 80g/kg dry matter, and a metabolizable energy value of about 12.5MJ/kg dry matter for cattle. The dried maize gluten feed is often made into pellets to facilitate handling. Since it is a milled product the fibre does not have the same effect as long roughage in ruminant diets. Nevertheless, maize gluten feed has been used as a substantial proportion of the concentrate feed of dairy cows.

Maize gluten meal can be used as a protein supplement for cattle but is more suitable for dry cows than milking cows because of its relatively low palatability. One useful feature of this product is that it has a high content of methionine. It may no longer be available or economic for feed use as it is being used extensively as a natural weedkiller in horticulture.

Wheat

Wheat is not traditionally used as a feed grain because its milling properties make it desirable for use as human food. Some wheat is grown, however, for feed purposes. In some situations wheat may be competitively priced with other feed grains. Feed-grade wheat is a palatable, digestible source of nutrients, which can be used in cattle diets if fed with caution to avoid digestive upsets. By-products of the flour milling industry are also very desirable ingredients for livestock diets.

The use of home-grown wheat in dairy cattle feeding has gained interest over the past few years for several reasons. As a result of the Common Agricultural Policy, the price of wheat has been lowered since the end of the 1990s. Furthermore, there is an increased interest among farmers in reducing feeding costs by growing cereal crops such as wheat.

Whole-crop wheat is a valuable alternative to grass silage in drier climates or in climates where other forages such as maize are difficult to grow. In climatic regions favourable for production of maize silage and grass or grass silage, home-grown wheat grain can partially substitute for commercial concentrates in cattle diets.

Wheat is close to maize in energy content but it contains more protein, therefore it can be used as a replacement for maize as a high-energy ingredient in supplementary grain mixtures and it requires less protein supplementation than maize. However, wheat is very variable in composition. The protein content, for example, may range from 60 to 220g/kg dry matter, though it is normally between 80 and 140g/kg dry matter. Therefore, periodic testing of batches of wheat for nutrient composition is recommended.

Wheat is low in fibre and high in starch content. It is higher in protein than maize, barley, or oats. Like all other cereal grains, wheat is deficient in calcium and adequate in phosphorus for cattle.

Processing wheat improves its digestibility. Due to its small kernel size, the increase in digestibility is substantial, 20–25% compared to an improvement in processed barley of 12–15%. According to research published in Australia, whole wheat has a digestibility of 60% compared with 86% for rolled wheat.

Wheat, especially if finely milled, forms a paste-like mass in the mouth and rumen and this may lead to digestive upsets. A general recommendation is that coarse rolling should be used to break the kernel into two or three pieces. Usually, wheat grain is fed dry and rolled or ground. Tempering wheat (by adding moisture) has been shown to be effective in reducing fines and maintaining milk yield and composition.

Wheat is very palatable if not ground too finely; good results having been obtained when wheat was coarsely ground (hammer mill screen size of 4.5–6.4mm).

According to research findings, up to 200g/kg steam-rolled wheat can be included successfully in the diet of dairy cows, provided adequate fibre is provided and the diets are properly formulated and mixed. An adaptation period of 2–3 weeks should always be provided, with an initial level not exceeding 100g/kg in the grain mix.

Wheat milling by-products are also useful feed supplements for ruminants. These by-products are usually classified according to their crude protein and crude fibre contents, and are traded under a variety of names that are often confusing, such as pollards, offals, shorts, wheatfeed and middlings.

Wheatfeed (middlings). Wheat middlings are generally very palatable and readily consumed by all classes of ruminants, with no additional processing. They are often an economically competitive source of protein or energy in cattle diets. The protein in wheat middlings is highly degraded in the rumen and well-utilized by cattle on low quality forages that are usually low in rumen-degradable protein. Mature forages are usually low in phosphorus and wheat middlings are a good source of this mineral and several trace minerals. Since they contain higher levels of fibre and lower levels of starch than whole wheat, digestive disturbances are less of a concern. However, an adjustment period to introduce cattle to this feedstuff is recommended. Wheat middlings can be an effective supplement for beef cows grazing low quality winter range or being fed low quality forages.

Sorghum

Sorghum, commonly called grain sorghum or milo, is the third most important cereal crop grown in the USA and the fifth most important cereal crop grown in the world. Grain sorghum has a hard seed coat, therefore grinders should be set to break all of the kernels without producing a fine, dusty feed.

Research findings indicate that sorghum and maize are somewhat similar in nutritive value as supplementary grains for cattle feeding. One disadvantage of grain sorghum is that it can be more variable in composition because of growing conditions. Crude protein content usually averages around 89g/kg, but can vary widely from 70 to 130g/kg, therefore a protein analysis prior to formulation of diets is recommended.

The hybrid yellow-endosperm varieties are more palatable to livestock than the darker brown sorghums, which possess a higher tannin content to deter wild birds from damaging the crop.

Sorghum in ground and processed forms has been used successfully in a range of cattle feeds. Several investigations have shown that heat processing improves the nutritive value of sorghum for dairy cattle, making it equal to steam-flaked maize. This is explained by a greater proportion of the starch being fermented in the rumen, enhanced digestibility of the reduced fraction of dietary starch reaching the small intestine, and increased total starch digestion.

Organic farmers would have difficulty steam-flaking sorghum on-farm, but might find it worthwhile to purchase the processed sorghum from a feed supplier or feed processor.

Triticale

Triticale is a hybrid of wheat and rye that combines the grain quality, productivity and disease resistance of wheat with the vigour, hardiness and high lysine content of rye. Globally, triticale is used primarily for livestock feed, its advantages over wheat and barley making it of interest to organic cattle producers. In nutrient content it is similar to wheat, with a digestibility similar to that of maize. As with other grains, processing by rolling, milling or crushing has been shown to improve the digestibility. Coarse-rolling the grain using a roller mill is preferable to fine milling with a hammer mill.

As with rye, triticale is subject to ergot infestation. Studies using this grain have shown an increased incidence of liver abscesses in steers when compared with sorghum diets. As a consequence, it is recommended that triticale be limited to a maximum of 50% of the grain portion of livestock diets.

Component g/kg unless stated	Barley	Maize	Oats	Sorghum	Triticale	Wheat	Wheat middlings
Dry matter	881	900	892	900	900	890	890
Energy							
DE (kcal/kg)	3840	3920	3400	3620	3410	3880	3310
DE (MJ/kg)	16.08	16.41	14.23	15.16	14.28	16.24	13.85
ME (kcal/kg)	3030	3250	2780	2960	3000	3180	2890
ME (MJ/kg)	12.68	13.61	11.64	12.39	12.56	13.26	12.1
NEm (kcal/kg)	2060	2240	1850	2000	1860	2180	1940
NEm (MJ/kg)	8.62	9.38	7.75	8.37	7.79	9.13	8.12
NEg (kcal/kg)	1400	1550	1220	1350	1230	1500	1290
NEg (MJ/kg)	5.86	6.49	5.11	5.65	5.15	6.28	5.4
NEl (kcal/kg)	2050	1960	1740	1830	1780	2040	1950
NEl (MJ/kg)	8.58	8.21	7.29	7.66	7.45	8.54	8.16
Fibre							
Crude fibre	34	23.0	120	28	33	28	88
NDF	181	108.0	293	180	220	118	370
ADF	87	33.0	140	40	50	126	118
Fat							
Crude fat (g/kg)	20.0	37.0	52	32	16.0	24	47
Protein							
Crude Protein	132	98.0	136	126	165	142.0	185
By-pass (%)	24	58	18	55	25	28	22
Minerals							
Ca	0.5	0.3	1.1	0.7	0.5	0.5	1.3
P	3.5	3.2	4.1	3.4	3.7	4.3	10.2

TABLE 5.1 Average composition of common cereal grains and grain products for organic ruminant feeding.*

* Air-dry basis. ** N × 6.25

OIL SEEDS, OIL FRUITS, THEIR PRODUCTS AND BY-PRODUCTS

The major supplementary protein sources used in animal production are oilseed meals. Only those meals resulting from mechanical extraction of the oil from the seed are acceptable for organic diets.

Soya beans, canola, groundnuts, and sunflowers are grown primarily for their seeds, which produce oils for human consumption and industrial use. Among these crops, soya beans comprise the predominant oilseed produced internationally. Another approved oilseed is cottonseed, grown as a source of textile and the oil used for food and other purposes.

Moderate heating is generally required to inactivate anti-nutritional factors present in oilseed meals. As a group, the oilseed meals are high in protein content, except safflower meal with hulls. The extent of dehulling affects the protein and fibre contents, whereas the efficiency of oil extraction influences the oil content and thus the energy content of the meal. Oilseed meals are generally low in calcium and higher in phosphorus. The amino acid composition is less important for ruminant animals than for pigs and poultry because of the way proteins are digested in ruminants.

Soya Beans and Soya Bean Products

Soya beans are grown mainly as a source of oil for the human market, a by-product being soya bean meal that is used widely in animal feeding. Whole soya beans are also being used in animal feeding.

Soya bean meal is generally regarded as one of the best plant protein sources in terms of its nutritional value. It also has a complementary relationship with cereal grains in meeting the amino acid requirements of farm animals. As a consequence, it is the standard by which other plant protein sources are compared.

Whole soya beans contain 150–210g/kg oil, most of which is removed in the oil extraction process. For use in organic feeding the oil has to be extracted using hydraulic or screw presses (expellers). The mechanical process is less efficient than the chemical solvent process in extracting the oil but results in advantages for cattle feeding. One is that the heat generated by friction of the screw presses inactivates the anti-nutritional factors present in raw soya beans. The heating also makes the protein less digestible, resulting in a higher content of rumen by-pass protein and improved milk production. Another benefit is that expeller soya bean meal has a higher content of residual oil than solvent-extracted meal, resulting in a higher energy value.

Whole soya beans contain 360–370g/kg protein, and the extracted meal 410–500g/kg protein. Studies cited by Blair (2011) have shown that soya bean meal has a good amino acid profile for ruminant feeding. For instance, Holstein bull calves weaned at 6 weeks of age were used in four experiments to identify the limiting amino acids in a maize/soya bean meal diet. The results indicated that methionine was the first-limiting amino acid, followed by lysine and tryptophan. Therefore, an appropriate combination of soya bean meal and cereal grains provides an excellent dietary protein mixture for all classes of ruminants, particularly calves.

The energy value of soya bean meal is higher than in most other oilseed meals, due mainly to its low fibre content.

Full-Fat Soya Beans

Whole soya beans have a higher energy value than soya bean meal and can be utilized effectively in feeding cattle, particularly when combined with low-energy ingredients. Generally the beans are rolled or ground, but this is not essential for use with adult cattle. Processing limits the storage life of soya beans due to increased potential for rancidity, therefore processed raw soya beans should be used within 7 days. Cooking or other high heat treatment will inactivate the enzymes that cause the fat to become rancid, allowing a potentially longer storage time. Also, because of possible rancidity problems, diets based on raw full-fat soya beans should be used immediately and not stored, unless an approved antioxidant is added to the diet.

The beans need to be heated in some manner when fed to calves to inactivate the anti-nutritional factors present. Roasting is a common process used, and has additional benefits in drying the beans, reducing possible mould contamination and increasing the content of rumen by-pass protein.

Some research has shown that feeding soya beans to beef cattle increased the amount of polyunsaturated fatty acids in the meat. This type of fat is now preferred to saturated fat in the human diet, therefore this may be a way of enhancing the nutritive value of beef and other meats.

Canola (Improved Rapeseed)

Canola was developed from industrial rapeseed by plant breeders in Canada during the 1960s, resulting in seed containing a food grade oil and an improved meal for animal feeding. Canola is a small diameter seed (1–2mm), with the hull making up a significant amount of the seed weight, about 16%.

Canola is now the main type of rapeseed being grown. Since 1991 virtually all rapeseed production in the EU has been of cultivars similar to canola, with a low content of erucic acid and a low content of glucosinolates (anti-nutritional factors).

Canola meal is produced from canola seed by oil extraction. As with soya bean meal, only expeller canola is acceptable as an organic feedstuff, the expeller process resulting in similar benefits.

Canola seed contains about 400g oil, 230g protein and 70g crude fibre/kg. The oil is high in polyunsaturated fatty acids, which makes it valuable for the human food market. It can also be used in animal feed. However, the oil is highly unstable due to its content of polyunsaturated fatty acids.

Canola meal is lower in crude protein than soya bean meal and varies according to cultivar. Depending on the processing method, the content of by-pass protein in canola meal is slightly lower or similar to that of soya bean meal. Owing to its higher fibre content (>110g/kg), canola meal contains less energy than soya bean meal and the fibre tends to be lower in digestibility. Compared with soya beans, canola seed is a good source of calcium, selenium and zinc, but a poorer source of potassium and copper.

Research findings indicate that the palatability of modern cultivars of canola is high and without the mustard-type flavour associated with unimproved rapeseed. This has allowed canola meal to be included at dietary levels up to 200 and 300g/kg for calves and dairy cows, respectively. No negative effects on milk or beef flavour have been reported.

Full-Fat Canola (Canola Seed)

A more recent approach with canola is to include the whole (unextracted seed) in cattle diets as a convenient way of providing both supplementary protein and energy.

Good results have been achieved with this feedstuff, especially with the low-glucosinolate cultivars. This could be a good use of the product by organic farmers who are able to grow the crop on-farm.

Whole canola seeds can be fed to adult cattle, preferably rolled or cracked prior to feeding. As with raw soya beans, canola seed should not be stored for more than 7 days after processing due to potential rancidity problems.

The seed needs to be processed frequently and stored for short periods only. Once ground, the oil in full-fat canola becomes highly susceptible to oxidation, resulting in undesirable odours and flavours. The seed contains a high level of α-tocopherol (vitamin E), a natural antioxidant, but additional supplementation with an acceptable antioxidant is needed if the ground product is to be stored. A practical approach to the rancidity problem is to grind just sufficient seed for immediate use.

Cottonseed

Whole cottonseed is a widely used feedstuff for dairy cattle because of its combination of high fibre, energy (from fat) and protein. For instance, one survey in 1999 showed that approximately 40% of dairy producers in the USA fed whole cottonseed.

Recent data from the USA showed an average concentration of 225 crude protein, 388 acid detergent fibre, 472 neutral detergent fibre, and 178g/kg fat in the dry matter. The composition is known to be affected by cultivar. Compared with other commonly available protein supplements, whole cottonseed is the only one that combines high energy and high fibre. This feature makes it attractive in feeding early lactation dairy cows that are likely to be in negative energy balance and with a reduced appetite. Controlled heating improves the protein

value of both whole cottonseed and cottonseed meal by causing a lower degradation in the rumen and a greater transfer of amino acids to the small intestine.

Cottonseed meal is the second most important protein feedstuff in the USA. As with other oilseeds, only mechanically extracted meal is acceptable in organic feeding. The protein content of cottonseed meal may vary from 360 to 450g/kg, depending on the contents of hull and residual oil. The fibre content is higher in cottonseed meal than soya bean meal, the energy value being inversely related to the fibre content.

Cattle and sheep are less susceptible than pigs to the presence of an anti-nutritional factor – gossypol – in cottonseed. This is a natural toxin present in the cotton plant that protects it from insect damage. Adult cattle have the ability to detoxify some gossypol in the rumen but this factor limits the levels of cottonseed or cottonseed meal that can be included in calf diets. Ingestion of excess gossypol can have a detrimental effect on fertility in both males and females. This problem does not occur with glandless cultivars of cottonseed. Calves should not be fed cottonseed until they have been weaned. The recommended upper safe level of whole cottonseed is 1.0kg per cow per day and 0.6kg per weaned calf per day.

Whole cottonseed and cottonseed meal can be used successfully used in beef cattle diets.

Groundnuts (Peanuts)

Whole groundnuts, which are about 10 to 20% shell by weight, typically contain 60–80g/kg moisture and, on a dry matter basis, 220–260g/kg crude protein and 360–4,400g/kg oil. Groundnuts not suitable for human consumption can be used as a protein and energy supplement for beef cows with access only to poor quality grass hay or poor quality pasture. Mature beef cattle will consume raw, whole groundnuts, including the shells. The groundnuts may have to be blended with a grain such as maize or barley before they are readily consumed by heifers.

Groundnut meal is high in protein (400–450g/kg) and can be a good substitute for other protein supplements such as soya bean meal and cottonseed meal. However, groundnuts are known to be susceptible to fungal infections before, during and after harvest. As a result, the groundnuts may be contaminated with toxic aflatoxin produced by the fungus *Aspergillus flavus*, causing health problems in a range of farm livestock, particularly young stock. Levels of 150–200g/kg of toxic groundnut in dairy cow diets have resulted in a significant reduction in milk yield. Feed standards in several countries now place strict limits on aflatoxin content of feeds, which has led to a reduction in the usage of groundnut meal.

Sunflower Seed

Sunflower is an oilseed crop of considerable potential for organic cattle production since it grows in many parts of the world. Sunflower seed surplus to processing needs and seed unsuitable for oil production may be available for feed use. On-farm processing of sunflower seed is being done in countries such as Austria.

The seeds contain approximately 380g/kg oil, 170g/kg protein and 159g/kg crude fibre and are a good source of dietary protein and fat. Due to the high oil content, intake should be limited so the total dietary fat does not exceed 5% of the dietary dry matter.

Limited findings indicate that whole sunflower seed can be fed to dairy cows as an alternative to other oilseeds and that it can be used without any processing of the seeds, such as roasting. Whole sunflower seeds can

also be fed to beef cattle without processing, cattle finding the seed highly palatable. There appears to be no advantage in cracking or rolling sunflower seeds prior to feeding. The size of the seed results in cattle chewing and breaking down the product during feeding. Including the sunflower seed in a mixed diet eliminates any issues of feed preference or palatability.

Sunflower meal is produced by extraction of the oil from sunflower seeds. The nutrient composition of the meal varies considerably, as a result of variation in the extent of dehulling and the efficiency of the oil extraction process. The crude fibre content of whole (hulled) sunflower meal is around 300g/kg and with a complete decortication (hull removal) the fibre content is around 120g/kg. Energy value varies substantially, related to fibre level and residual oil content. Higher levels of hulls included in the final meal lower the energy content and reduce the bulk density. The mechanical process of oil extraction leaves more residual oil in the meal, often 50–60g/kg. The higher oil content in mechanically extracted meals makes this product a useful feed supplement for dairy cows, since lactating cows often respond to supplementation of the diet with fat. The energy value of the meal is lower than that of canola or soya bean meal, with an ME value of about 13 MJ/kg dry matter for cattle.

Sunflower meal does not have a good amino acid profile for growing stock and is not recommended for young calves or lambs, but it can be utilized successfully by older animals.

Sunflower plants can be made into silage, which can be utilised effectively by older livestock. Immature sunflowers result in silage with a higher nutrient content. Generally, sunflower silage contains a higher content of crude protein and fat than maize silage, but also a higher content of crude fibre and lignin. As a result, the energy content is lower, indicating that sunflower silage should be fed to lower producing dairy cows, dry cows, or growing heifers.

LEGUME SEEDS, THEIR PRODUCT AND BY-PRODUCTS

Of those listed in this section of approved products, the main products used in the supplementary feeding of ruminant animals are field peas, field and broad beans and lupin seed.

Field Peas

Field peas are now used widely in animal feeding in several countries. They are of particular interest to organic farmers since they can be grown and used on-farm. Peas are a good cool season alternative crop for regions not suitable for growing soya beans.

There are green and yellow varieties, which are similar in nutrient content. Those grown in North America and Europe, both green and yellow, are derived from white-flowered varieties. Brown peas are derived from coloured flower varieties. They have higher tannin levels, lower starch, higher protein and higher fibre contents than green and yellow peas. These varietal differences account for much of the reported variation in nutrient content.

Peas are similar in energy content to high energy grains such as maize, but they have a higher protein content (about 230g/kg). Peas contain a high level of starch, which is highly digestible. The starch type is similar to that in cereal grains and the starch content has been found to be correlated inversely with protein content. Feed peas, like cereal grains, are low in calcium but contain a slightly higher level of phosphorus (about 4g/kg). The levels of trace minerals and vitamins in peas are similar to those found in cereal grains.

Field Beans

Field (faba) beans grow well in regions with mild winters and adequate summer rainfall and the beans store well for use on-farm. Like peas, they are commonly regarded nutritionally as high-protein cereal grains. They contain about 240 to 300g/kg protein and an energy value similar to that of barley. The crude fibre content is around 80g/kg, air-dry basis. The oil content of the bean is relatively low (10g/kg dry matter), but with a high proportion of linoleic and linolenic acids that make the beans very susceptible to rancidity if stored for more than about 1 week after grinding. When fresh they are very palatable. As with the main cereal grains, faba beans are a relatively poor source of calcium and are low in iron and manganese. The phosphorus content is higher than in canola meal. Like peas, they should be ground before being included in diets for ruminant animals.

Lupins

Lupins are becoming of increasing importance as a feed ingredient. Australia is the world's leading producer and exporter of lupin grains, representing 80–85% of the world's production and 90–95% of the world's exports.

Lupins are a valued component of cereal cropping rotations in Australia, especially across large areas of Western Australia. Benefits of this crop for the organic producer are that the plant is a nitrogen-fixing legume and, like peas and faba beans, can be grown and utilized on-farm with minimal processing. Another advantage of

Component	Canola meal	Cottonseed meal	Faba beans	Field peas	Groundnut meal	Soya bean meal	Lupin seed sweet white	Sunflower seed meal
Dry matter (g/kg)	937	926	870	891	924	900	890	920
Energy								
DE (kcal/kg)	3380	3480		3820	3870	3760		
DE (MJ/kg)	14.15	14.57		16	16.2	15.74		
ME (kcal/kg)	2960	2630		3410	2780	3340	3344	2240
ME (MJ/kg)	12.39	11.01		14.28	11.64	14	14	9.4
NEm (kcal/kg)	1840	1720		2140	1850	2100		
NEm (MJ/kg)	7.7	7.2		8.96	7.74	8.79		
NEg (kcal/kg)	1210	1110		1470	1380	1430		
NEg (MJ/kg)	5.07	4.64		6.15	5.78	6		
NEl (kcal/kg)	1760	1810		2000	1920	1980		1380
NEl (MJ/kg)	7.37	7.58		8.37	8.04	8.29		
Fibre								
CF (g/kg)	129	129	81	63	67	650		137
NDF (g/kg)	269	280	151	139	140	220	223	400
ADF (g/kg)	204	180	112	79	61		183	
Fat								
Crude fat g/kg	101	50	15.4	10.0	55	80	107	10
Protein								
CP (g/kg)	352	443	278	263	342	420	387	280
By-pass (%)	30	50		22	69	35	25	16
Minerals (g/kg)								
Ca	7.2	2.0	1.2	1.4	2.2	3.6	3.7	4.3
P	12.6	11.7	5.2	4.6	6.0	6.6	4.1	11.5

TABLE 5.2 Average composition of common protein feedstuffs for organic ruminant feeding.*

* Air-dry basis. ** N × 6.25

lupins is that the seed stores well. The shortage of organic protein feedstuffs in Europe has stimulated interest there in lupins as an alternative protein source for cattle and other ruminants.

The development of low-alkaloid (sweet) cultivars has allowed the seed to be used as animal feed. In Australia, where much of the research on lupins as a feedstuff has been carried out, the main species of lupins used in animal feed are *L. angustifolius, L. luteus* and *L. albus*.

Lupin seed has a thick seed coat, resulting in a crude fibre content of 130–150g/kg in *L. luteus* and *L. angustifolius* and a slightly lower content in *L. albus*. The crude fibre component contains more hemicellulose than other legumes such as peas and faba beans, which have cellulose as the major component.

The carbohydrate profile of lupins is different from that of most legumes, with negligible levels of starch and high levels of soluble and insoluble non-starch polysaccharides (mainly galactans) and oligosaccharides. These features influence the utilization of energy from lupins and may explain the range in energy values reported for lupins. The crude protein content of *L. angustifolius* has been reported as ranging from 272–372g/kg and of *L. albus* from 291–403g/kg (air-dry basis), with some having protein levels comparable to soya bean meal if dehulled.

The oil content of lupins appears to vary within and between species and is generally quite low, and with a high level of antioxidant activity in the seed that helps to explain the good storage characteristics of lupins. Lupins are low in most minerals, with the exception of manganese.

TUBER ROOTS, THEIR PRODUCTS AND BY-PRODUCTS

Among these feedstuffs approved for inclusion in organic diets are sugar beet pulp, dried beet, potato, sweet potato as tuber, manioc (cassava or tapioca) as roots, potatoes, potato pulp (by-product of the extraction of potato starch), potato starch, and potato protein. Although sugar beet pulp and dried beet are listed in Table 2.1, other root crops are not listed, therefore their acceptability in organic diets should be confirmed with the local organic certifying agency.

The most important root crops used in ruminant feeding are potatoes, turnips, swedes (rutabagas), mangels (mangolds) and fodder beet. Two by-products of the sugar extraction industry, sugar beet pulp

and molasses (from both sugar beet and sugarcane), are important and nutritionally valuable animal feeds.

Tubers differ from root crops in containing either starch or fructan instead of sucrose or glucose as the main storage carbohydrate. They have higher dry matter and lower fibre contents than root crops.

The main characteristics of root crops are a high moisture content (750–940g/kg) and a low crude fibre content (40–130g/kg dry matter). The organic matter consists mainly of sugars (500–750g/kg, dry matter basis) and is of high digestibility. Roots are generally low in protein content, although like most other crops this component can be influenced by the application of nitrogenous fertilizers. The composition also varies with size, large roots having lower dry matter and fibre contents and of higher digestibility than small roots. Winter hardiness is associated with higher dry matter content and keeping quality.

Potatoes

As with other root crops, the major drawback is the relatively low dry matter content (180 to 250g per kg) and consequent low nutrient density. Potatoes are variable in composition, depending on variety, soil type, and growing and storage conditions. On a dry matter basis, whole potatoes contain about 60 to 120g per kg crude protein, 2 to 6g per kg crude fat, 20 to 50g per kg crude fibre and 40 to 70g per kg ash. About 70% of the dry matter is starch and the crude protein content is similar to that of maize, so that potatoes can be regarded as a cereal replacement. The fibre and mineral contents are low (with the exception of potassium).

Potatoes may contain the glycoside solanin, particularly if the potatoes are green and sprouted, and may result in gastroenteritis and toxicity. Consequently, such potatoes should be avoided for feeding. The water used for cooking should be discarded and not fed to animals because it may contain the water soluble solanin. Ensiling also destroys some of the toxin so that inclusion of slightly greened potatoes with grass should be acceptable.

There is a lack of recent research on potatoes for cattle. However, potatoes have an established history of usage in dairy cow and beef cattle feeds as a cereal replacement. The available evidence indicates that cattle should be introduced gradually to potatoes in the diet. One advisory note regarding the feeding of potatoes is that choking may occur in cattle due to the ingestion of whole tubers.

Potatoes can be ensiled and the silage is accepted readily by ruminant animals. It is similar to maize silage in nutrient content.

Sweet Potatoes

The sweet potato is a very important tropical plant whose tubers are widely grown for human consumption and as a commercial source of starch. The reddish cultivars are often called yams.

The tubers are of similar nutritional value to ordinary potatoes although of much higher dry matter and lower crude protein contents. Sweet potatoes contain about 300g per kg dry matter, mostly starch with some sugars. The crude protein content is around 50–70g per kg (dry matter basis). The main challenge relating to the incorporation of sweet potatoes in cattle diets is the need for additional protein to counter the low crude protein content of this crop.

As sweet potatoes do not keep as well as potatoes they are sometimes cut into slices and dried. Problems of sweet potato poisoning have been reported in cattle, attributed to fungal infection that causes pneumonia.

These reports indicate that only sound, undamaged tubers of sweet potatoes, or dried product made from them, should be fed to cattle. Also the storage conditions should prevent mould development.

Cassava

Cassava is an approved ingredient in organic cattle diets, although in many countries it will represent an imported product, not produced regionally.

Fresh cassava contains about 650g per kg moisture. The dry matter portion is high in starch and low in protein (20 to 30g per kg, of which only about 50% is in the form of true protein). It has a high crude fibre content (about 270g/kg dry matter basis) because of the presence of the peel. Cassava can be fed fresh, cooked, ensiled, or as either dried chips or (usually) as dried meal. One way of avoiding the powdery nature of ground cassava is to use dried cassava chips.

Cassava is an excellent energy source because of its content of highly digestible carbohydrates (700–800g/kg), mainly in the form of starch. Its energy value is similar to that of potatoes. Cassava can be used to partly replace cereal grain in cattle diets provided the diet can be formulated to adjust for the negligible amount of protein and micronutrients provided by cassava.

Fresh cassava contains cyanogenic glucosides (mainly linamarin), which on ingestion are hydrolyzed to hydrocyanic acid and are poisonous to animals. Boiling, roasting, soaking, ensiling, or sun-drying are used in the countries where cassava is produced, to reduce the levels of these compounds. The normal range of cyanide in fresh cassava is about 15 to 500 mg/kg and a general recommendation is that the level of HCN equivalent (HCN, linamarin and cyanohydrins combined) must not exceed 50mg per kg in the complete feed.

Cassava has been found to be useful as a source of energy in supplementary diets for pastured cattle in Africa and other tropical areas.

An example of the results found with cassava in an importing country was provided in a study conducted by Brigstocke et al. (1981) in the UK. The study involved the inclusion of cassava in a concentrate feed provided to Friesian cows along with grass silage. The concentrate contained cassava at 0 or 400g/kg and 600 and 103g per kg barley, respectively. Average daily intake of silage and of concentrate were not affected significantly by inclusion of cassava. Average daily milk yield, without or with cassava, was 21.14 and 22.27kg and milk fat content was 41.4 and 40.4, respectively. These differences and differences in the content of milk solids were not significant statistically.

Turnips, Swedes

Turnips are often used to extend the grazing season or to provide supplementary feed for winter feeding. Much of the relevant research has been done in Australia and New Zealand.

Swedes (*Brassica napus*) and turnips (*Brassica campestris*) have a similar composition, although turnips generally contain less dry matter and a lower energy value than swedes. Yellow-fleshed turnips are of higher dry matter content than the white-fleshed varieties. These crops can be considered as a replacement for cereal grains.

Research in Switzerland cited by Blair (2011) showed that turnips had an average yield of 6.5 tonnes per hectare dry matter, higher than a grass–clover ley (2.9 tonnes/ha) or other brassica species. Grazing losses were 33% on average, giving a net yield of 4.3 tonnes per hectare dry matter. In spite of treading damage, no indications of long-term impacts on the soil were found. There

was no negative influence on animal health, milk quality or milk taste. However, other research has shown that feeding turnips in place of grain can alter milk composition, in particular by lowering the fat content and the content of unsaturated fatty acids. As a consequence, SFC (an index of milk fat hardness) also increased.

Both swedes and turnips are liable to taint milk if given to dairy cows around milking time. The volatile compound responsible for the taint is absorbed from the air by the milk and is not absorbed from the feed during digestion. However, this effect has not been reported consistently with turnip feeding.

Mangels, Fodder Beet and Sugar Beet

These are all members of the same species, *Beta vulgaris*, and for convenience they are generally classified according to their dry matter content. Mangels are the lowest in dry matter content, highest in protein and lowest in sugar content of the three types. Fodder beet can be regarded as being between mangels and sugar beet in terms of dry matter and sugar content, while sugar beet is highest in dry matter, sugar and energy contents, though lowest in protein content.

In the past, root crops have been considered as an alternative to silage in ruminant diets, but their value as cereal replacements is now recognized.

Fodder beet is a poor source of protein but is popular in Denmark and some other European countries as an energy source for dairy cattle and young ruminants. In addition to varietal type, the dry matter content of fodder beet is influenced by the stage of growth at harvesting and environmental conditions. Research results suggest that dairy cows should be introduced gradually to fodder beet. Care is required in feeding cattle on high dry matter fodder beet since excessive intakes may cause digestive upsets, hypocalcaemia and even death. The digestive disturbances are attributed to the high sugar content of the beet.

It is customary to store mangels for a few weeks after lifting, since freshly lifted mangels may have a slightly purgative effect. The toxic effect is associated with the nitrate content, which in storage is converted into asparagine. Unlike turnips and swedes, mangels do not cause milk taints when fed to dairy cows.

Sugar Beet

Most sugar beet is grown for commercial sugar production, though it is sometimes fed to cattle. Owing to its hardness, the beet should be pulped or chopped before feeding.

After extraction of the sugar at a sugar beet factory, two valuable by-products are obtained, which are used as feed ingredients in cattle feed: sugar beet pulp and beet molasses.

Sugar Beet Pulp

Most sugar beet pulp is now sold after drying and the addition of molasses. The crude fibre content is relatively high (about 200g/kg dry matter) and the crude protein content is low at about 100g/kg dry matter. Beet pulp contains high amounts of pectins that can reduce the risk of ruminal disorders compared with feedstuffs high in starch. Addition of molasses raises the water soluble carbohydrate (i.e. sugar) content from 200 to 300g/kg DM. Molassed sugar beet pulp is used commonly as a feed for dairy cows.

Beet Molasses

This product of sugar production contains about 700–750g/kg dry matter (mainly sugars) and is difficult to handle as a feed component. The crude protein content is low and is mainly in the form of non-protein nitrogenous compounds. Molasses can be used to improve the palatability of feeds for calves, etc., but only at low levels since it is laxative. Other uses are as a binding agent to reduce dustiness in the feed and as a binding agent in cubes, pellets and the compressed feed blocks that are used as protein, mineral and vitamin supplements for ruminants. Beet molasses is also used as an additive in ensilage.

Cane Molasses

This product is similar to beet molasses and is used as a feed ingredient in tropical countries where the crop is grown. In some countries such as Cuba it is used as a supplementary energy source in forage-based diet and as the main ingredient of beef cattle diets, providing 500–800g/kg of total dry matter intake. While use of such diets has given acceptable live weight gains, instances of a condition known as molasses toxicity have been reported. This is characterized by inco-ordination and blindness caused by deterioration of the brain, attributed to an unusual rumen fermentation pattern resulting in VFA mixtures high in butyrate and low in propionate. The condition appears to be avoided by ensuring that the cattle have access to good quality forage.

Other Feed Sources

Cabbages have a high yield of nutrients per hectare and are of potential interest for organic feeding as a source of roughage. However, little research appears to have been conducted on this crop as a feed ingredient for cattle. Data cited by Blair (2011) indicate that cabbage (variety Drumhead)

Component	Roots and tubers/by-products						
	Cassava (manioc)	Potatoes	Swedes	Beet pulp dried	Molasses cane	Molasses beet	Seaweed meal
Dry matter (g/kg)	880	210	103	910	710	770	930
Energy							
DE (kcal/kg)		3490		3100	3090	3380	
DE (MJ/kg)		14.61		13	12.94	14.15	
ME (kcal/kg)		3080		2680	2670	2940	
ME (MJ/kg)		12.9		11.22	11.18	12.31	
NEm		1920		1760	1640	1980	
NEm		8.04		7.37	6.87	8.29	
NEg		1280		1140	1030	1333	
NEg		5.36		4.77	4.31	5.58	
NEl		1820		1700	1600	1830	
NEl		7.61		7.12	7	7.66	
Fibre							
CF (g/kg)	49	20	120	210	0	0	70
NDF (g/kg)	93	40	440	460	0	0	
ADF (g/kg)	68	30	340	250	0	0	110
Fat							
Crude fat (g/kg)	55	40	18	7.0	43	2.0	33
Protein							
CP (g/kg) **	33	100	120	100	60	80	70
By-pass (%)		0	0	44		0	
Minerals (g/kg)							
Ca	2.4	0.7	5.5	9.1	12.4	1.4	27.2
P	2	3.0	6.8	0.8	1.2	0.3	3.1

TABLE 5.3 Average composition of common root and tuber crops for organic ruminant feeding.*

* Air-dry basis. ** N × 6.25

contained 100g dry matter per kg and (per kg, dry matter basis) 18MJ gross energy, 230g crude protein, 79g true protein, 7.6g total lysine, 4.7g methionine + cystine, 142g acid detergent fibre and 132g ash.

All the feedstuffs mentioned above should be drawn from the lists of approved organic feedstuffs detailed in Chapter 2, or otherwise meet organic standards.

NUTRIENT REQUIREMENTS

The nutrient requirements of ruminant livestock have been assessed by authorities in several countries, based on research findings related to conventional production. Unfortunately no comparable publications on organically raised livestock have been published to date. Nevertheless, organic livestock require the same nutrients as conventional stock, therefore it is possible to use these estimates of requirement in formulating recommendations for the feeding of organic cattle, sheep and goats.

In addressing a similar issue with the feeding of pigs and poultry, Blair (2007, 2008) recommended that the established requirements for these species be modified from the values that had been derived from research involving modern, fast-growing hybrid stock. The rationale for the modification was that organic production favours the use of heritage, pure-bred stock that grows more slowly and has a lower productivity than hybrid stock. The modification suggested dietary mixtures containing essential nutrients at a lower concentration, more in keeping with the level of production expected in organic stock.

It is questionable whether in the case of organic ruminants that such a modification is necessary. One reason is that there is much less of a difference in the type of stock used in organic and conventional production than in pig or poultry production. Both organic and conventional ruminant production favour the use of pure-breds mainly, with a lower usage of hybrids and cross-breeds. Therefore the established requirements for conventional ruminants are likely to have been derived from stock more similar to the stock used in organic production, than in the case of organic pigs and poultry. In spite of this, productivity is generally lower in organic cattle than in conventionally raised cattle. The most likely reason for this is the high level of forage mandated in organic cattle diets. Therefore the difference in productivity is due to diet rather than genetic make-up of the animals.

Another reason relates to environmental issues. A common perception is that pasture-based, low-input dairy systems characteristic of the 1940s are more conducive to environmental stewardship than modern milk production systems. This has been tested by modelling the typical management practices, herd population dynamics and production data from US dairy farms (Blair, 2011). Modern dairy practices required considerably fewer resources than dairying in 1944 with 21% of animals, 23% of feedstuffs, 35% of the water and only 10% of the land required to produce the same 1 billion kg of milk. Waste outputs were similarly reduced, with modern dairy systems producing 24% of the manure, 43% of the CH^4 (methane) and 56% of the N^2O (nitrous oxide) per billion kg of milk compared to equivalent milk from historical dairying. The carbon 'footprint' per billion kg of milk produced in 2007 was 37% that of equivalent milk production in 1944. The conclusion of the study was that to fulfil the increasing US requirement for dairy products it is essential to adopt management practices and technologies that improve productive efficiency to allow milk production to be increased, while reducing resource use and mitigating environmental impact.

An important consideration that is emerging in organic cattle production is its influence on greenhouse gas emissions, particularly methane. This gas is considered to have 21 times the global warming potential of carbon dioxide. It has been estimated that beef production worldwide accounts for about 62% of total livestock methane emissions, milk 19%, sheep 12%, pigs 5% and poultry 1%. Estimates suggest that livestock in Asia and the Pacific produce 33% of total methane emissions, Latin America 23%, Europe 14%, Africa 14%, North America 11% and Oceania 5%.

As explained above in the section on digestion of carbohydrates, fibrous diets promote a higher production of methane than more readily digested diets. This explains why organic milk production inherently increases methane emission, unless the animals are fed highly digestible diets. Similarly, organic beef cattle emit more methane than conventional beef cattle.

These factors, i.e. the type of stock used and the environmental issues (in particular the greenhouse gas emission potential), suggest that the most recent estimates of requirement should be used as the basis for feeding organic cattle. This allows us to take advantage of all relevant knowledge relating to the mitigation of methane emissions. The high level of forage mandated in the diet of organic cattle does make these requirement values more difficult to attain in practice than with conventional cattle, and requires that the forages used be of high quality. The benefit for the organic cattle industry is then that low quality forages are discouraged, with a reduction in the accompanying problem of greenhouse gas emissions.

Nutrient requirement tables used in North America and in several other countries are based on the recommendations of the National Research Council (NRC), National Academy of Sciences (NAS), Washington, DC., the most recent publications being the Nutrient Requirements of Beef Cattle (NAS–NRC, 2000: an update is expected to be published in 2016), the Nutrient Requirements of Dairy Cattle (NAS–NRC, 2001) and the Nutrient Requirements of Small Ruminants (2007).

Examples of the tabulated requirements are shown in tables 5.4 and 5.5.

The 2001 NAS–NRC publication *Nutrient Requirements of Dairy Cattle* is much more complex and comprehensive than previous issues. A very useful feature is that a feed ration can be formulated using a CD-ROM issued with the book. The programme allows environmental factors such as temperature, wind speed and distance walked to be taken into account, so that the feed requirements can be tailored more exactly to the particular situation in question. *Nutrient Requirements of Beef Cattle* (NAS–NRC, 2000) also contains a CD-ROM for ration formulation.

A key question is whether the range of data used by the NAS–NRC to generate the mathematical models includes values likely to be encountered in herds of organic cattle. Correspondence between the author of this book and the NAS–NRC committees indicates that such is the case.

Some organic producers prefer to use the earlier (1989) versions of the NAS–NRC *Nutrient Requirements of Dairy Cattle* since it is less complex. An example of the requirement data contained in this publication is shown in Table 5.5.

FEEDING ORGANIC DAIRY CATTLE

Replacement Stock

Newborn calves are best left with the mother for a minimum of 4 days, to obtain an adequate intake of colostrum and get a healthy start. The initial health of the calves depends largely on how well the cows were fed during the last 2 months of gestation.

The calves can then stay with the mother or be moved to a nurse cow that will accept several calves. Older cows are generally more experienced and less likely to reject calves. Should supplementary milk be required, as judged by the behaviour and appearance of the calves, whole milk from other cows should be fed or milk reconstituted from an acceptable source of milk powder. This supplementary milk should be fed over several feedings, preferably from a container equipped with a drinking nipple. Calves fed from an artificial teat tend not to suck on each other or on objects, unlike calves fed from a bucket.

After a week, calf starter or creep feed (160g/kg protein) can be offered in a creep

Body wt. kg	200	250	300	350	400	450
Maintenance						
NEm, Mcal/d	4.1	4.84	5.55	6.23	6.89	7.52
MP g/d	202	239	274	307	340	371
Ca g/d	6	8	9	11	12	14
P g/d	5	6	7	8	10	11
Growth kg/d	NEg Mcal/d					
0.5	1.27	1.50	1.72	1.93	2.14	2.33
1.0	2.72	3.21	3.68	4.13	4.57	4.99
1.5	4.24	5.01	5.74	6.45	7.13	7.79
2.0	5.81	6.87	7.88	8.84	9.77	10.68
2.5	7.42	8.78	10.06	11.29	12.48	13.64
	MP required for gain, g/d					
0.5	154	155	158	157	145	133
1.0	299	300	303	298	272	246
1.5	441	440	442	432	391	352
2.0	580	577	577	561	505	451
2.5	718	712	710	687	616	547
	Calcium g/d					
0.5	14	13	12	11	10	9
1.0	27	25	23	21	19	17
1.5	39	36	33	30	27	25
2.0	52	47	43	39	35	32
2.5	64	59	53	48	43	38
	Phosphorus g/d					
0.5	6	5	5	4	4	4
1.0	11	10	9	8	8	7
1.5	16	15	13	12	11	10
2.0	21	19	18	16	14	13
2.5	26	24	22	19	17	15

TABLE 5.4 Nutrient Requirements of Growing and Finishing Cattle (from NAS–NRC, 2000). (Example Angus cattle weighing 200–450kg and gaining 0.5–2.5kg per day).

feeder. It should be provided several times a day, allowing the calves to eat as much as they want. The feeders should be cleaned out after the calves have had a reasonable amount of time to feed. Free access to leafy, high quality legume hay should also be provided.

Research by Weary (2001) has provided valuable information for the organic dairy farmer on the correct rearing of calves. His research showed that calves do very well when kept with the cows during the first few weeks after birth, gaining weight at up to 3 times the rate of conventionally reared calves (i.e. separated early and fed milk at 10% of body weight per day). He pointed out that under natural conditions cows leave their calves in groups from about 2 weeks of

Live weight kg	Energy				Total CP g	Minerals	
	NEI MCal	ME MCal	DE MCal	TDN kg		Ca g	P g
Maintenance of mature lactating cows							
400	7.16	12.01	13.80	3.13	318	16	11
450	7.82	13.12	15.08	3.42	341	18	13
500	8.46	14.20	16.32	3.70	364	20	14
Maintenance plus last 2 months of gestation of mature dry cows							
400	9.3	15.26	18.23	4.15	875	26	16
450	10.16	16.66	19.91	4.53	928	30	18
500	11.0	18.04	21.55	4.9	978	33	20
Milk production, nutrients per kg milk of different fat percentages							
Fat %							
3.0	0.64	1.07	1.23	0.28	78	2.73	1.68
3.5	0.69	1.15	1.33	0.301	84	2.97	1.83
4.0	0.74	1.24	1.42	0.322	90	3.21	1.98
4.5	0.78	1.32	1.51	0.343	96	3.45	2.13
5.0	0.83	1.4	1.61	0.364	101	3.69	2.28
5.5	0.88	1.48	1.7	0.385	107	3.93	2.43

TABLE 5.5 Daily Nutrient Requirements of Lactating and Pregnant Dairy Cows (from NAS–NRC, 1989).

FIGURE 5.3 Cattle type suitable for organic farming.

age and usually continue to nurse calves for more than 6 months. In a number of organic milk production systems, the heifer calves suckle the dam for 4 days (Denmark) to 8 weeks (Sweden). Producers report healthier and faster growing calves and believe that this management system reduces the incidence of mastitis.

Another finding was that calves can easily consume 9 or more litres of milk a day, compared with the 4 litres they receive when fed conventionally. The increased milk intake greatly improved weight gains, with no detrimental effect on calf health or post-weaning intake of solid feed. The calves can be reared successfully in small groups without stimulating cross-sucking or increasing the incidence of disease. Weary (2001) also found that cows kept with calves yielded less milk at milking. However, this was attributed to a lack of milk ejection at milking and not to reduced milk synthesis. Yields rebounded after separation such that total yield over the lactation period did not differ.

Another issue investigated by Weary (2001) was the belief that calves should be encouraged to increase their consumption of starter feed at an early age. He found that over the first 5 weeks of life, feeding calves less milk did increase starter consumption (0.17 versus 0.09kg per day) but that this practice severely limited weight gains. Moreover, he found that calves fed milk to appetite quickly caught up to conventionally fed calves in their intake of starter after weaning. Both groups consumed on average 1.9kg per day during the 2 weeks after weaning.

Examples of feed mixtures for use with calves from about one week of age are shown in Table 5.6. The feed should be introduced

Ingredient g per kg air-dry basis	1	2	3	4	5	6
Maize grain rolled	500	390	540	500	340	280
Ear maize ground					140	
Oats rolled	350		120	260	340	300
Barley rolled		390				
Wheat bran		100	110			
Soya bean meal expeller	130	100	80	170	160	150
Linseed meal			80			
Beet pulp						200
Molasses, liquid			50	50		50
Dicalcium phosphate	10	10	10	10	10	10
Trace mineral, salt & vitamin mix	10	10	10	10	10	10
Calculated analysis as-fed basis						
Crude protein, g/kg	145	140	145	154	147	148
NEm, MJ/kg	7.66	7.37	7.54	7.66	7.03	7.33
NEg, MJ/kg	5.23	4.98	5.11	5.23	4.65	4.98
Calcium, g/kg	2.9	2.9	3.5	3.4	3.2	4.5
Phosphorus, g/kg	5.4	6.1	6.4	5.4	5.2	4.9
Dry matter, g/kg	885	884	878	878	889	885

TABLE 5.6 Example feed mixtures for dairy calves (from Chiba, 2009).

gradually to stimulate rumen development and allow continued live weight gain after weaning. Good quality hay or forage should also be provided from about 10 days of age, although appreciable quantities will not be consumed until the calves are about 8 to 10 weeks of age.

Mixtures 1–4 are suggested for calves weaned after 4 weeks of age and receiving forage. Mixtures 5 and 6 are suggested for calves weaned after 4 weeks of age and not receiving forage.

Grower feed (Table 5.7) can be introduced to heifers at around 4 months of age. Although these animals will be ruminating by this stage it is likely that the rumen capacity is insufficient to allow all the required nutrients to be provided from pasture. In addition, pasture-reared heifers have to expend high levels of energy for maintenance due to their increased activity and exposure to climatic conditions that are often less than ideal. As a result, the amount of energy available for growth may be limiting and supplementation required.

Mixture 1 is designed for feeding with legume hay; mixtures 2 and 3 are designed for feeding with legume-grass hay; and mixture 4 is designed for feeding with grass hay. As with other feed mixtures, alternative feedstuffs can be formulated into the feed mixture to provide a similar content of energy and nutrients.

The amount of supplement to be provided should be based on the observed growth rate and condition of the heifers in relation to those of the breed and strain of animal in question. The allowance can then be adjusted upwards or downwards accordingly. The aim is to have the heifers achieve adequate growth rates without becoming too fat.

Young bulls can be fed similarly to heifers except that they grow faster and consequently require more feed. Mature bulls can be maintained mainly on forage with minimal feeding of supplement.

Dairy Cows

It is clear from the above that dairy cows on pasture or fed forage are likely to need supplementary feed, at least during some stages of their reproductive cycle. A main question is how to calculate the amount and composition of supplement needed. With pigs and poultry it is usually possible to use standard formulas for feed mixtures based on average nutrient content of feedstuffs. Appropriate feedstuffs, minerals and vitamins can then be purchased to supplement the home-grown grains. Similar standard formulas are available for use with dairy cows, but are

Ingredient g per kg air-dry basis	1	2	3	4
Maize grain cracked	780			500
Ear maize ground			760	
Oats rolled	200	350		270
Barley rolled		500		
Soya bean meal expeller		80	170	200
Molasses, liquid		50	50	
Ground limestone				10
Dicalcium phosphate	10	10	10	10
Trace mineral, salt & vitamin mix	10	10	10	10
Calculated analysis as-fed basis				
Crude protein, g/kg	92	138	139	167
NEm, MJ/kg	7.83	7.16	7.70	7.62
NEg, MJ/kg	5.4	4.86	5.32	5.23
Calcium, g/kg	2.5	3.3	3.5	6.8
Phosphorus, g/kg	4.8	5.6	4.9	5.6
Dry matter, g/kg	879	884	867	886

TABLE 5.7 Example feed mixtures for dairy heifers (from Chiba, 2009).

	High protein		Medium protein		Low protein	
Ingredient g per kg air-dry basis	Example 1	Example 2	Example 1	Example 2	Example 1	Example 2
Maize grain		700			500	
Ear maize ground	920		740	780		610
Oats ground or rolled		280				
Wheat bran					230	
Molasses, liquid						60
Soya bean meal	60			200	240	300
Soya beans, cracked			240			
Dicalcium phosphate	10	10	10	10	10	10
Ground limestone					10	10
Trace mineral, salt & vitamin mix	10	10	10	10	10	10
Calculated analysis as-fed basis						
Crude protein, g/kg	99	95	152	152	189	187
TDN, g/kg	714	742	735	717	716	705
NEL, MJ/kg	1.65	1.72	1.70	1.65	1.66	1.63
Calcium, g/kg	2.9	2.5	3.4	3.2	7.0	7.6
Phosphorus, g/kg	4.5	4.8	5.1	5.1	7.6	5.5
Dry matter, g/kg	869	881	881	874	886	871
Calculated analysis dry matter basis						
Crude protein, g/kg	114	108	172	174	213	214
TDN, g/kg	822	842	834	820	808	809
NEL, MJ/kg	1.90	1.95	1.93	1.89	1.87	1.87
Calcium, g/kg	3.3	2.8	3.8	3.7	7.9	8.7
Phosphorus, g/kg	5.2	5.4	5.8	5.8	8.6	6.3

TABLE 5.8 Example supplementary feed mixtures for dairy cows fed forage of high, medium or low protein content (Chiba, 2009).

not advised for general use because of variability in the nutritive quality of the forage. Examples are shown in Table 5.8.

In order to formulate an exact supplementary feed mixture it is necessary to match the nutrients in the feed supply to the requirements of the cow. This cannot be done for each individual animal, therefore the cows should be managed in groups at the same stages of lactation and reproductive cycle.

An important first step is to have the forage analyzed for nutrient content. Forages and roughages vary greatly in nutrient content, and this variation is too great to allow the use of standard feed mixtures. Also, since forages constitute a very high proportion of the diet any errors would be magnified. Hence there is a need for laboratory analysis of a representative sample of the forage prior to its use. This allows the nature and extent of supplementation to be calculated.

Important information provided by laboratory testing includes the fibre and protein values. Cherney et al. (2009) advised that NDF (neutral detergent fibre) content is the most useful measure of quality. According to these authors there is a relatively small range in optimal NDF for lactating dairy cows but that there is as yet no reliable method of estimating the fibre content of grass and alfalfa–grass mixtures for use in timing harvesting operations. NDF has to be measured after the forage has been harvested and stored, prior to ration balancing. The optimum content of NDF appears to be 380g/kg for lucerne and 500g/kg for grass. These authors showed that milk production decreased linearly as the dietary forage content increased from 500 to 800g/kg. NDF intake remained constant as forage content increased from 50 to 80%, suggesting that when forage source is constant, NDF intake is a reliable predictor of dry matter intake and milk production. Further information can be derived from NDF. Knowledge of the NDF and DM values allows the NEL and TDN contents to be predicted, using standard equations. NEg values can be calculated in a similar way for use in beef cattle feeding.

Kersbergen (2010a) provided further information on the value of forage analysis reports. He advised that close attention should be paid to both acid detergent fibre (ADF) and neutral detergent fibre (NDF) levels. ADF allows a prediction of the available energy of the forage and NDF allows a prediction of the intake. Forages should represent 60 to 100% of the cow's diet to maintain rumen health and function, conveniently the range mandated for organic cattle feeding. He advised that a cow can

usually eat 0.8 to 1% of her body weight in NDF if the quality of the forage is poor, whereas she can eat up to 1.2% of her body weight in NDF if the forage is of high quality. On well-managed pastures that percentage can go even higher (1.4% of body weight in NDF). Quality forages can allow dairy cows to consume the equivalent of 3.5 to 4% of their body weight on a dry matter basis.

Table 5.9 shows an example of an organic forage analysis report, indicating some analytical parameters that producers can use as goals (Kersbergen, 2010b). These include crude protein at 232, NDF at 377 and ADF at 277g per kg (dry matter basis).

The dry matter intake predicted by the computer programme was 15.1kg per day, and the actual intake was 15.4kg per day. The energy and metabolizable protein provided by the ration would allow a milk yield up to 20.5–21.0kg per day, slightly higher than the actual yield measured before the ration was formulated.

The feeding system above was based on the cows having access to the outdoors but without access to grazing. A more typical situation on organic farms is for the cows to be grazing, at least during suitable weather conditions. The NAS–NRC (2001) computerized rationing programme provides the following ration for grazing Jersey cows at the same stage as those in the above example (Table 5.11).

The pasture intake in the above example is equivalent to 72.3% of total intake on a dry matter basis.

Several software programmes are available to allow farmers to run feed formulation programmes on their home computers. Alternatively, feed companies may supply this service.

It may seem out of place for organic farmers to be advised to use computing systems to obtain feed mixtures for their livestock. However, if the result is a more

◆ Dairy One

```
FORAGE TESTING LABORATORY       ------------------------- ---- ---- --------
DAIRY ONE, INC.                 |Sample Description    |Farm |Code| Sample |
730 WARREN ROAD                 |MMG SILAGE            |     |302 |11850910|
ITHACA, NEW YORK 14850          |-------------------------- ---- ---- --------
607-257-1272    (fax 607-257-1350) |
                                |-------------------------- ---- ---- --------
-------- -------- -------- -- -- |           Analysis Results            |
|Sampled | Recvd |Printed |BY|CO| |---------------------------- -------- --------|
|        |11/15/07|11/19/07|  | | |       Components      | As Fed |   DM   |
                                |-------------------------- -------- --------
     06 2008 3RD G/A             |% Dry Matter          | 46.9 |        |
UNIV OF MAINE CO-OP EXTENSION    |% Neutral Detergent Fiber| 17.7 | 37.7 |
EXTENSION CROPS TEAM            |% Crude Protein       | 10.9 | 23.2 |
LIBBY HALL                      |Soluble Protein % CP  |      |  62  |
ORONO, MI 04469                 |ADICP % CP            |      |  4.8 |
                                |% Crude Fat           |  2.0 |  4.3 |
-------------------------        |% Ash                 | 4.77 | 10.15|
       ENERGY TABLE - NRC 2001   |% Calcium             |  .57 | 1.22 |
BW = 1350 Pat% = 3.7 fprot% = 3.1|% Phosphorus          |  .17 |  .35 |
-------------------------        |% Magnesium           |  .14 |  .31 |
Milk,     NEL     NEL   Milk,    |% Potassium           | 1.40 | 2.99 |
Lb       Mcal/Lb Mcal/Kg  Kg     |% Sulfur              |  .11 |  .24 |
-------  -------  -------  ----   |% Sodium              | .025 | .054 |
Dry       0.72    1.59   Dry     |PPM Iron              |  50  | 106  |
40        0.69    1.52   18      |PPM Zinc              |  16  |  35  |
60        0.66    1.46   27      |PPM Copper            |   4  |   8  |
80        0.63    1.39   36      |PPM Manganese         |  11  |  23  |
100       0.59    1.31   45      |                      |      |      |
120+      0.55    1.21   54+     |% Acid Detergent Fiber| 13.0 | 27.7 |
-------------------------        |% ADICP               |  .5  |  1.1 |
NEM3X     0.70    1.53           |% NFC                 | 13.6 | 29.0 |
NE G3X    0.43    0.94           |% TDN                 |  30  |  63  |
ME 1X     1.15    2.53           |NEL, Mcal/Lb          |  .32 |  .67 |
DE 1X     1.34    2.95           |NEM, Mcal/Lb          |  .30 |  .64 |
TDN1X,%     63                   |NEG, Mcal/Lb          |  .18 |  .38 |
-------------------------        |Relative Feed Value   |      | 166  |
                                |                      |      |      |
COMMENTS:                        |% Moisture            | 53.1 |      |
1.NRC ENERGIES - SMALL BREEDS -  |% Available Protein   | 10.3 | 22.0 |
  DO NOT USE ENERGIES BEYOND 80  |% Adjusted Crude Protein| 10.9 | 23.2 |
  LBS. MILK. LARGE BREEDS - USE  |PPM Molybdenum        |  .9  |  1.9 |
  120 LB. ENERGY WITH EXTREME
  CAUTION.
```

As can be seen in the analysis report, the forage is of high quality. The reported crude protein level indicates that it contains some leguminous material. In addition to the analytical results the report also provides information on the predicted NEm ,NEg and NEl values at various levels of production, these values being predicted from the NAS–NRC (2001) equations.

With that information, together with breed/age information, a complete ration can then be formulated.

An example ration based on MMG silage (mixed mainly grass) was formulated for a mature Jersey cow (Kersbergen, 2010b), using the following specifications:

Animal type: lactating dairy cow	Breed: Jersey	Age: 37 months
Empty bodyweight: 450kg	Days pregnant: 15	Condition score: 2.60
Age at first calving: 22 months	Calving interval; 13 months	
Milk production: 20kg per day	Milk fat: 45g per kg	
Milk true protein: 32g per kg	Current temperature: 16°C	

The diet formulated, based on above forage and animal specifications, is as follows.

	DM kg per day	As-fed kg per day
MMG silage (see above)	10.60	22.60
Barley grain, ground	3.03	3.44
Soya beans, roasted	1.51	1.68
Mineral/vitamin supplement	0.25	0.25

The diet obtained, using the Cornell Net Carbohydrate and Protein System, is an acceptable feed mixture that could be fed as a total mixed ration. It contains almost 700g/kg forage. The nutrients provided by the diet are as follows, showing a slight surplus of energy, metabolizable protein, methionine, lysine, calcium, phosphorus and potassium. A small safety margin in terms of nutrients is preferred to a deficiency which would limit production or affect body condition.

TABLE 5.9 Example of organic forage analysis conducted by a forage testing laboratory in New York State, USA (Kersbergen, 2010b).

Requirement	ME Mcal/day	MP g/day	Met g/day	Lys g/day	Ca g/day	P g/day	K g/day
Maintenance	13.86	569	11	35	0	0	0
Pregnancy	0.03	1	0	0	0	0	0
Lactation	24.35	985	17	59	29	20	30
Growth	0	0	0	0	0	0	0
Total Required	38.25	1555	28	94	43	36	87
Total Supplied	39.49	1578	30	110	162	50	305
Balance	1.25	23	2	15	119	14	218

TABLE 5.10 Nutrients provided in silage-based ration formulated for mature Jersey cows (Kersbergen, 2010b)

	Dry matter kg/day	As-fed kg/day
Grass pasture	15.3	76.4
Barley grain, ground	2.80	3.1
Soya beans, roasted	2.21	2.43
Mineral/vitamin supplement	0.1	0.1
Legume forage hay	0.76	0.91

TABLE 5.11 Ration formulated as in Table 5.10 for cows grazing similar forage.

efficient use of resources, improved animal health, and a reduction in methane emission from the farm, then their use is justified. Also, computers are already used by many farmers for other uses such as bookkeeping.

One important point about ration development is that the nutrient requirements of dairy cows are not static but vary with stage of lactation. The feed mixture to be used at any stage has to be based, therefore, on the nutrients required during that stage. At peak production the cow may require 3–10 times as much protein and energy as in late gestation. A complication is that the voluntary intake (appetite) of the cow at peak production may be less than the intake necessary to fulfil the requirements. Maximum dry matter intake is not reached until 12 to 15 weeks after calving. The requirement for protein increases greatly at the start of lactation because milk contains about 32g per kg protein. The dietary protein, in addition to being adequate in amount, should also provide an optimal ratio of ruminally degradable protein to ruminally undegradable (by-pass) protein. The recommended ratio for high producing cows is around 60:40 (in the Jersey ration shown above it was 57:43).

It is also important that the diet of dairy cows provide sufficient calcium and phosphorus because of the high content of these minerals in milk. In the case of a Jersey cow, the recommended intakes are about 50–65g (absorbable Ca) per day and 35–55g (absorbable P) per day, respectively.

The reproductive cycle can be considered in various phases with differing nutrient requirements. The cows can then be fed appropriately for each phase.

The first phase is usually regarded as the first 6–10 weeks after calving. During this time intake is lower than optimal and peak milk production is reached. The cows respond by using body stores to make up for deficits in nutrient intake. The second phase is the period from 6–14 weeks after calving when intake is optimal and nutrient needs are in balance with the supply. The third phase is the remainder of the lactation period when intake exceeds requirements, the excess being used to build up body reserves for the next lactation.

Two main strategies are used by organic farmers to design an appropriate feeding plan for use during lactation. *Challenge feeding* is introduced in phase one. This involves giving each cow, regardless of yield, an estimated allowance of the supplement formulated for the forage being used. The allowance is then adjusted up or down according to the production of each cow. The strategy is continued in the second phase, when each cow is fed to match her measured milk yield. This strategy means feeding each cow individually and is more difficult to implement in large herds unless automated equipment is available. However, it can result in feed savings and a reduction in the risk of fat cows from over-feeding.

Phase feeding is the other approach that can be taken. In early lactation (phase one), high quality supplement mixtures are fed. Later in phase two, these high quality supplements are replaced by lower quality supplements.

Challenge feeding is probably the simpler procedure for organic farmers to adopt.

During the dry period the aim should be to ensure that each cow is in good condition for the next calving, but not too fat. Good forage may provide all of the required nutrients during this phase, but a supplement may be required in the final 3 to 4 months of gestation.

Cows are designed to graze forage, therefore it is usually recommended that feed troughs be placed so that the cows eat in a body position similar to that when grazing on pasture. Cows eating with their heads down produce more saliva, which increases their ability to buffer the rumen from excess acidity. The general recommendation is that the feed trough should be 10 to 15cm higher than the surface on which the cows are standing.

Feeding during the dry period may require that feed be restricted to avoid the cows becoming too fat, resulting in metabolic disturbances during the early part of lactation. However, a gradual increase in supplement intake is generally recommended during the final 6–8 weeks before calving. This is to adapt the cows to a higher intake of feed at the start of lactation and minimize or avoid the negative energy balance that occurs at that time.

A similar approach is generally recommended for bred heifers, to allow for growth to mature size without the animal becoming too fat.

Beef Cattle

As in organic dairying, the aim of organic beef farming is to optimize the available resources on the farm rather than to maximize the output of meat. Forage is the main feed of beef cattle and much of the requirement for energy and protein can be provided by rangeland and pasture. Hay and silage can be used when weather conditions restrict or prevent grazing.

The organic regulations require that at least 60% of the feed dry matter must be supplied by forage produced on the farm itself. Some organic producers feed even higher levels, in some cases providing forage as the sole feed. Such a level may be suitable for low-production stock provided it is

Output	Production system	
Animal basis	Organic	Conventional
Daily live weight gain, kg	0.84	0.86
Age at slaughter, months	17.5	17.1
Weight at slaughter, kg	499	497
Carcass weight, kg	267	268
Hectare basis		
Stocking rate, number	3.42	4.46
Live weight gain, kg	1481	1921

TABLE 5.12 Comparison of the growth performance of Hereford × Friesian steers in organic and conventional (intensive) 18-month beef systems (Younie and Mackie, 1996).

supplemented with necessary minerals and vitamins.

In addition, the proportion of concentrates in the diet is restricted to 40% on a daily dry matter basis. Low quality pastures can be used but, in view of their influence on methane production in the animals, the quality should be improved as much as possible.

The objective with breeding herds of organic beef animals is a high production of strong calves, therefore calving at the start of the forage growing season is the preferred system. The peak requirement for nutrients is then matched by the availability of high quality forage.

Younie and Mackie (1996) showed that efficient beef production can be obtained in an organic system, provided the forage quality is good (Table 5.12).

Factors influencing the choice of breed include the size of the breeding herd and the length of the grazing season (Younie, 2001). For instance in small herds in northern Europe it is common to use a pure-bred dam such as Angus and to adopt a pure pedigree breeding policy. In this environment a cow of small to medium mature size is better able to maintain itself on grass alone and to cope with a relatively short grazing season. This fits with the organic philosophy

of selecting breeds well adapted to the environmental conditions of the farm. Ease of calving, satisfactory temperament, adequate milk production, and the production of a premium beef calf are also important traits, as is natural polling.

Other breeds may meet many or possibly all of these objectives but the Angus has performed well in these conditions (Younie, 2001). In addition to its genetic suitability to a grass-based organic system, the Aberdeen–Angus breed has an additional high quality image that complements the organic brand image.

Once the herd has achieved a target size of 200 cows, a terminal sire of another breed such as Hereford or Simmental may be selected for breeding to poorer females (Younie, 2001). Cross-bred cows benefit from hybrid vigour, particularly in enhanced milk production. On some farms in northeast Scotland a March–April calving programme has been adopted to take maximum advantage of the seasonal cycle. Calving in spring has several important advantages for organic beef production. These are (Younie, 2001):

- For a herd wintered outside, spring calving ensures that all animals carried into winter are approximately 6 months old at least and better able to stand up to harsh weather

- Cows approach calving at the period of the year when they are at their leanest, thus reducing problems of dystocia
- Incidence of excess milk production is reduced
- Fly problems are reduced or non-existent, reducing the incidence of mastitis
- Calving can take place outdoors with reduced risk of severe weather (although with a greater risk than summer calving)
- Cows take advantage of spring grass peak growth rates at their lowest body condition and during an increasing demand for milk for the calf. This makes them very efficient biologically
- Peak milk yield on grass alone is achieved at the height of grass productivity, thus maximizing annual lactation yield
- Fertility is maximized by a rising plane of nutrition
- Calves begin grazing when grass productivity, quality and palatability are high
- Cows can lay down significant body reserves during the latter part of the grazing season, which can assist with body heat insulation and in reducing winter feed requirements
- Weaning takes place around late December to late January, leaving an independent calf well able to cope with a forage diet
- Weaned calves approach their second grazing season with good frames and moderate to thin body condition, again allowing maximum use of grazing. This allows the sale of finished animals to take place from pasture alone in September–November.

Establishing a good grass cover in spring, and withholding stock from grazing until the sward is ahead of future demand, pays very large dividends. At one of the university farms described by Younie (2001) in that area, stock is turned out directly on to their main summer grazing at a fairly heavy stocking rate of about 3.5 cows per hectare in May or June. There is no spring grazing of silage fields and the first cut is taken in late June, followed by a second cut in mid-August. All cows and calves are housed in late October, when the cows start to receive silage and mineral supplements, plus straw. Calves receive a daily creep feed of 500g organic cereal, mineral supplement and seaweed meal. After weaning, the feeding rate of cereal is increased to 1kg per head per day. In their second (finishing) winter, the animals are fed silage to appetite plus 2 to 3kg per day of organic cereal plus a mineral supplement. For finishing heifers, cereal feeding may be delayed or reduced in order to avoid finishing too early at light carcass weights or at high fat cover. No purchased protein feedstuffs are fed to either cow, calf or finishing animal. Total cereal consumption per head is 120kg for calves in their first winter and 325kg for finishing animals.

Feeding the Cow Herd

It is convenient to divide the reproductive production cycle into four periods.

First trimester of pregnancy: Nutrients are needed for maintenance (plus lactation if the cow has a calf). Factors influencing the requirements include breed, body weight, milk yield and milk composition. Body condition should be monitored and supplementary feed provided when required. The cow usually nurses a calf throughout this trimester. The trimester ends when the calf is weaned. Creep feeding is commonly practised with conventional production but is less common in organic production and is advised if the cows are not producing enough milk and calf growth is inadequate. One benefit of creep feeding (apart from improved calf growth) is that the cows are more likely to re-breed more quickly and require less feed to rebuild body reserves. Another is that the calf crop is more likely to be uniform in weight and that post-weaning weight loss is more likely to be reduced following creep feeding.

Creep feeding usually starts when the calves are about 3 weeks old. The creep feeder should be located close to water, shade, and a salt box. Only a small amount of creep feed should be placed in the feeder until the calves start to eat. This is to ensure freshness.

Heifers that are to be kept as cow replacements are usually not given creep feed.

Grazing cattle should have access to a mineral supplement, in the form of loose mineral or a block. A high level of salt or other substance can be included to limit consumption, if acceptable under the local organic regulations.

Second trimester: The calf is weaned and lactation ends. This is the period of lowest nutrient needs. Body condition should continue to be monitored and supplementary feed provided if required. This is the period when it is easier to make adjustments to the cow's body condition. Body condition scoring (BCS) can be used to assess the amount of energy reserves in the form of fat and muscle of beef cows. The scores used in North America range from 1 to 9, with a score of 1 being extremely thin and 9 being very obese. Areas such as the back, tail head, pins, hooks, ribs, and brisket of beef cattle can be used to determine BCS. Cows should calve with a BCS score between 5 and 7. Ideally, cows should achieve this score by the end of the second trimester and be managed to maintain it throughout the third trimester.

Third trimester: Nutrient needs are increasing rapidly due to foetal growth. Body condition has to be monitored to prevent cows becoming too fat/thin. Failure to meet the cow's nutrient needs at this time may result in a reduced percentage calf crop and it is possible that some cows may not produce a live, healthy, calf. An inadequate nutrient intake may also result in a cow that has difficulty re-breeding.

In a spring calving programme, the third trimester coincides with the winter season. Cold weather will increase the energy needs, and supplementary feeding of hay or silage may be required. First-calf heifers will be lower in social dominance and are likely to receive an inadequate amount of supplement unless fed separately from the cows. The third trimester ends at calving time.

Post-calving period: Lactation needs are high at this time and the reproductive system is recovering from calving. Adequate forage supplies need to be available at this time. Peak milk production is around 5–12kg per day, depending on breed, and the lactation period is over 100 days. Feed intake is 35 to 50% higher than in non-lactating animals. Good pasture together with a mineral supplement (free choice mixture of 50% dicalcium phosphate and 50% iodized salt) should provide all the required nutrients. If pasture quality is poor a supplement of silage or hay is beneficial. Lucerne hay or silage is a good choice when the protein content of the forage is low.

It is important to achieve conception in cows by 80 days post-calving in order to maintain a 12-month calving interval. Cows calving with a BCS of less than 4 are likely have a delay in onset of first oestrus following calving, increasing the time required for re-breeding. Cows, and particularly heifers, receiving inadequate nutrition during pregnancy have poor reproductive performance overall. On the other hand, cows calving with a BCS over 7 are likely to have a reduced conception rate.

The forage should be managed so that it

is of high quality. Mature forages and low quality hays are likely to result in a protein deficiency and poor reproduction. The ration should contain more than 70g crude protein per kg. An adequate water supply needs also to be provided.

It follows from the above that the most important period in terms of nutrition is from 30 days before calving until 70 days after calving. Cows gaining weight just before and during the breeding season show a shorter period between calving and first oestrus, and tend to have high conception rates.

In North America it is common to have either a spring (March through April) or autumn (September through October) calving. This avoids periods of very hot or very cold weather. The feeding programme can then be devised accordingly. Most producers favour spring calving, for the reasons outlined above by Younie (2001).

To achieve these goals, lactating beef cows need to receive sufficient nutrients in order to provide adequate milk to support calf growth. If the feed resources are inadequate, e.g. pasture quality is poor, the cow may be unable to produced sufficient milk for good calf growth. In this case, creep feed may be required for the calves. The composition could be similar to that described above by Younie (2001) or provided as hay, cracked grain or a mixed feed of 900g per kg grain, 50g per kg molasses and 50g per kg of a protein feedstuff such as soya bean meal.

Weaning of the calves from their dams generally takes place when the calves are 6 to 9 months of age.

Breeding Herd Replacements

Nutritional management of replacement beef heifers was reviewed. It is recommended that heifers to be kept as herd replacements be fed and managed separately from the cows. This is to ensure they receive adequate amounts of feed and grow uniformly to reach puberty in time to breed at 13 to 15 months of age. Typically puberty is reached when the animals reach 60% of their mature body weight (dual-purpose breeds reach puberty slightly earlier, at 55% of their mature body weight). Calving is then at 22 to 25 months of age. In addition to the nutrient demands of pregnancy, heifers require nutrients for growth so that they attain 85% of mature body weight at calving. Where the forage is deficient in certain trace elements, the deficiency should be corrected by providing the necessary nutrients in the form of a supplement, feed blocks or mineral licks, as with dairy herds.

Research conducted at the Agriculture Canada Research Station, Lethbridge, over an 8-year period indicated that the feeding of high- versus medium-energy diets to young British breed beef bulls (Hereford and Angus) was detrimental to their breeding ability. The high-energy diets consisted of 80% supplement (barley 600; oats 100; beet pulp 100) and 20% forage (alfalfa or alfalfa–straw (70:30) cubes), while the medium-energy diet was forage alone. Bulls were fed either high- or medium-energy diets from weaning until slaughter at 12, 15 or 24 months of age. At slaughter, sperm production by the bulls was estimated by epididymal sperm reserves. In most cases, regardless of age, bulls fed high-energy diets had substantially reduced reproductive potential compared with bulls fed medium-energy diets. Along with a reduction in sperm reserves, quality of semen and the libido of bulls fed high-energy diets were reduced. In addition to the detrimental influence on reproductive traits, it would be expected that bulls fed high-energy diets would have a much greater probability of acquiring foot and leg problems and consequently reduced longevity.

Very little comparable research has been conducted on bulls of Continental breeds. Results of a study conducted at Kansas State University showed no effect on either seminal characteristics or breeding capacity of feeding three different levels of energy to Hereford and Simmental bulls from weaning for a period of 200 days followed by grazing for 38 days. However, it should be noted that Hereford bulls in this study fed the lowest of three levels of energy had backfat thicknesses similar to bulls of the same breed and comparable age fed the high-energy diet referred to earlier in the Lethbridge Research Station study. A 3-year field trial conducted by the Lethbridge Research Station was designed to assess the effectiveness of different criteria used to evaluate the reproductive capacity of young beef bulls used for multiple-sire natural service under range conditions. Numerous measurements, including backfat thickness, were taken immediately before the breeding season. A total of 277 bulls representing five composite 'breeds' of cross-bred bulls were included in the analysis. Bulls were composed of Brown Swiss, Charolais, Chianina, Geibvieh, Limousin, Romanola, and Simmental breeds. Bull fertility was determined by blood-typing calves to determine their paternal parent. The average backfat thickness of all bulls was 1.5 + 0.07mm (range 0 to 7mm). There was a highly significant negative contribution of backfat thickness to bull fertility, indicating that as backfat thickness increased, bull fertility decreased. As a result of this finding it is now recommended that cattle ranchers select bulls with minimum backfat thickness, in order to optimize reproductive capacity.

Market Animals

Weaned calves not kept as breeding herd replacements are grown to market weight as meat animals. The simplest system is to pasture the cattle during the grazing season and to feed them preserved forage during winter. Rotational grazing systems can be used to maximize grass production and minimize infection from internal parasites. After weaning it is common for organic cattle at this stage to receive some cereal grain such as oats or barley over the winter period (e.g. 0.5 increasing to 1kg per head per day) in addition to silage or other preserved forage. In their second (finishing) winter, cattle that again receive cereal grain and a mineral supplement in addition to silage fed to appetite are more likely to achieve high grades at slaughter than when forage-feeding only is allowed. Generally animals fed exclusively on forage have to be taken to a higher market weight to achieve good grades.

SHEEP

Grazing is the basis of an organic sheep production system. This type of operation generally involves spring lambing, pasturing throughout the summer, marketing of lambs in the autumn either as finished lambs or lambs for further feeding, breeding of ewes in early winter, and wintering on pasture with appropriate shelter provided. The sheep graze all year, with supplemental feeding provided in those periods when forage is in short supply.

During the grazing season, sheep are able to meet their nutritional requirements from good pasture and a mineral supplement. Pastures of mixed grass and clover, lucerne and root crops such as turnips provide excellent sources of nutrition for sheep. A source of clean, fresh water should be provided at all times.

Permanent pasture should be the predominant source of nutrition for the sheep

FIGURE 5.3 Sheep type suitable for organic farming.

flock. It is advisable for clover to be over-seeded on pastures in winter to improve the quantity and quality of forage produced during the grazing season. Sheep prefer to graze leafy, vegetative growth that is 5–15cm tall rather than stemmy, more mature forages.

Pasture growth does not occur evenly throughout the year. Approximately 60% of the annual dry matter production of most species of cool season grasses occurs in the spring. When pastures are not stocked heavily enough to use the spring flush of growth, sheep graze and re-graze certain areas while other areas are left to mature and go to seed. This type of grazing behaviour weakens those plants that are grazed more frequently and gives the less desirable plants a competitive advantage.

Some of the spring pasture can be fenced off for hay production. After a hay cut, the pasture should be given a 3- to 4-week recovery period before making it available for grazing over the rest of the year. It is advisable to use a rotational grazing programme designed to move the sheep every 10 to 14 days, starting in early summer in order to improve both pasture and lamb production. More intensive rotational grazing systems with higher stocking rates can be used to promote more complete forage utilization, but are likely to result in higher levels of internal parasitism and reduced lamb growth.

Pasture for sheep can include any mixture of grasses, legumes, brush and shrubs. The carrying capacity of the farm depends on factors such as soil type, plant species, amount of precipitation, temperature and topography of the land. A general rule of thumb is 6 sheep per hectare of high quality pasture, but less if control of internal parasites is difficult under the prevailing organic regulations.

Sheep will consume a wide variety of forages, and selectively graze several weeds such as multiflora rose and blackberry. Companion grazing of sheep with other species of livestock, such as cattle or goats, results in greater pasture utilization and higher quality pastures than when a single species is grazed alone. Sheep prefer to graze hillsides and steep slopes and they provide a means of improving forage utilization and fertility in areas difficult to access with farm equipment.

The feeding strategy should be to match stage of production with type and quality of forage available. Lactating ewes with lambs should have access to the highest quality pasture available to promote satisfactory levels of milk production and lamb growth. Dry, non-pregnant ewes or ewes in early to mid-gestation can be placed on lower quality forages or follow-up as second grazers behind young, growing lambs.

Supplementation with grain and protein mixtures, together with vitamins and minerals, is likely to be needed when the ewes are pregnant or milking. Supplementation may also be required in finishing lambs for market, although sheep are excellent converters of forage to meat and fibre and have the ability to produce excellent carcasses from forage alone.

Sheep require a relatively high content of protein in the diet because protein is needed for wool production. Supplementary protein can be provided as cereal grains, legume hay, field peas, or canola, soya bean or sunflower meal.

The most recent official publication on the nutrient requirements of sheep is the *Nutrient Requirements of Small Ruminants: Sheep, Goats, Cervids, and New World Camelids* (NAS–NRC, 2007), which can be used as the basis of feed formulation. However, many organic producers prefer to use the previous (NAS–NRC, 1985) tabulation of requirements since it is less complex and is more user-friendly. Examples of the tabulated requirements in this publication are shown in Tables 5.13, 5.14 and 5.15.

As advised by Umberger (2016), ewe body weight does not remain constant throughout the year, but fluctuates with stage of production. Nutrient requirements are lowest for ewes during the maintenance period, increase gradually from early to late gestation, and are highest during lactation.

The feeding strategy is determined by assessing the ewe body weight and body condition score at three stages of production: 1) 3 weeks before breeding; 2) mid-gestation; and 3) weaning. Condition score is determined by handling ewes down their back, with a score of 0 indicating an extremely thin animal and 5 indicating a very fat animal. This is the easiest and best method of monitoring the nutritional status and overall well-being of the breeding flock. Ideally, ewes should range from a condition score of 2.5 at weaning to 3.5 at lambing. When necessary, thin ewes are separated

Body wt. kg	Daily gain g	Dry matter intake kg	Intake as % body wt.	TDN kg	Crude protein g	Calcium g	Phosphorus g
50	10	1.0	2.0	0.55	95	2.0	1.8
60	10	1.1	1.8	0.61	104	2.3	2.1
70	10	1.2	1.7	0.66	113	2.5	2.4
80	10	1.3	1.6	0.72	122	2.7	2.8
90	10	1.4	1.5	0.78	131	2.9	3.1

TABLE 5.13 Daily Nutrient Requirements of Ewes for Maintenance (from NAS–NRC, 1985).

Body wt. kg	Daily gain g	Dry matter intake kg	Intake as % body wt	TDN kg	Crude protein g	Calcium g	Phosphorus g
50	100	1.6	3.2	0.94	150	5.3	2.6
60	100	1.7	2.8	1.0	157	5.5	2.9
70	100	1.8	2.6	1.06	164	5.7	3.2
80	100	1.9	2.4	1.12	171	5.9	3.6
90	100	2.0	2.2	1.18	177	6.1	3.9

TABLE 5.14 Daily Nutrient Requirements of Breeding Ewes (from NAS–NRC, 1985). Flushing: 2 weeks pre-breeding and first 3 weeks from breeding.

Body wt. kg	Daily gain g	Dry matter intake kg	Intake as % body wt	TDN kg	Crude protein g	Calcium g	Phosphorus g
50	−60	2.4	4.8	1.56	389	10.5	7.3
60	−60	2.6	4.3	1.69	405	10.7	7.7
70	−60	2.8	4.0	1.82	420	11.0	8.1
80	−60	3.0	3.8	1.95	435	11.2	8.6
90	−60	3.2	3.6	2.08	450	11.4	9.0

TABLE 5.15 Daily Nutrient Requirements of Breeding Ewes (from NAS–NRC, 1985), first 8 weeks of lactation, suckling twins.

and fed additional energy to increase body condition. Conversely, obese ewes are separated and fed a lower energy diet at a stage of production when body weight loss is acceptable. Generally, problems with over-fat ewes are fewer than those with ewes that are too thin.

For 2 weeks before breeding and continuing 2 weeks into the breeding period, ewes should be placed on high quality pasture or supplemented daily with about 0.5–0.75kg whole grain. This practice is termed flushing and has been shown to improve lambing percentage by 10 to 20%. Flushing works best with mature ewes that are in moderate body condition, and has been shown to be more effective for early and out of season breeding than at the seasonal peak of ovulation during the autumn/early winter. Most prenatal deaths occur within the first 25 days after breeding and are usually associated with poor nutrition. Therefore, it is

important to avoid drastic reductions in nutrient supply during the breeding season.

Pastures with more than 50% clover or other legumes should be avoided during the breeding season because legumes may contain oestrogenic compounds that reduce conception rates.

From breeding to 6 weeks before lambing, the ewe flock can be maintained on permanent pastures, small grain pastures, conserved forage, aftermath crop fields, or hay. Foetal growth is minimal, and the total feed requirement of the ewe is not significantly different from a maintenance diet at this time.

The developing foetus makes most of its growth during the last 6 weeks of pregnancy. Therefore, it is important to supplement ewes with about 300–400g of whole grain in addition to their normal diet starting 6 weeks before lambing, to prevent pregnancy toxemia, low birth weights, weak lambs at birth, and low milk production. However, producers should be careful not to over-feed grain during late gestation, which could result in lambing difficulty caused by large lambs. The amount offered depends on the condition of the ewes and quality of the forage. The forage portion of the ration should provide all the protein required. It may be useful to separate the ewes into groups according to age, condition, and number of foetuses and be fed accordingly.

After lambing, the energy and protein requirements of the ewe increase due to the demands of milk production. Protein supplementation is especially important for ewe flocks with a high percentage of multiple births. Unless high quality pasture or legume hays are available, protein supplementation will likely be required in addition to grain. A general rule of thumb for the supplementary feeding of lactating ewes is 250g per day of grain for each lamb nursing the ewe. Ewes should be sorted into feeding groups based on type of rearing (single, twin, etc.) to ensure that grain supplements are not over- or under-fed.

Examples of rations formulated to meet the requirements of an 80kg ewe at different

stages of production and with determined energy and crude protein values for the hay available are shown in Table 5.16.

The period from weaning to re-breeding of ewes is critical if a high twinning rate is to be achieved. Ewes should not be allowed to become excessively fat but should make daily gains from weaning to breeding. A good target is that live weight should be around 60%–70% of projected mature weight at breeding and 80%–90% of projected mature weight at lambing. If pasture production is inadequate, ewes should be fed high quality hay and a small amount of grain if necessary. After mating, ewes can be maintained on pasture. Good pasture for this period allows the ewes to enter the winter feeding period in good condition.

Feeding Lambs

Lambs born at the end of winter are placed on pasture with their dams, where they remain throughout the spring and summer. They should have access to a high quality creep feed by the time they are 7 days old. The creep feed should contain 180–200g/

		Protein content of hay g/kg			
		165	150	125	100
Stage of production	Feedstuff				
Maintenance	Hay kg	1.5	1.5	1.5	1.5
Early pregnancy	Hay kg	1.6	1.6	1.6	1.6
Late pregnancy	Hay kg	1.9	1.9	1.9	1.9
	Maize kg	0.5	0.5	0.5	0.5
	SBM g				50
Early lactation (twins)	Hay kg	2.25	2.25	2.25	2.25
	Maize kg	1.0	0.9	0.85	0.5
	SBM g		50	200	350

TABLE 5.16 Amounts of hay, shelled maize, and soya bean meal (SBM) required to meet the total digestible nutrients (TDN) and crude protein (CP) requirements of a 80kg ewe when the hay contains different protein levels (from Umberger, 2016).

Feed ingredient kg	Example 1	Example 2	Example 3
Hay			460
Barley			520
Maize	600	885	
Oats	285		
Canola meal			110
Soya bean meal	100	100	
Ground limestone	10	10	5
Trace-mineralized salt	5	5	5
Total	1000	1000	1000

TABLE 5.17 Examples of feed mixtures for growing lambs (air-dry basis).

kg crude protein and be high in energy. Usually creep feed is composed of mixtures of ground grain and protein meals. Alternatively it can be fed as pellets. A general guide in feeding lambs is to allow 5–6 cm of trough space per lamb and to provide fresh water in the creep area.

A 2:1 calcium to phosphorus ratio is a good guideline for on-farm feed mixes. A ratio of less than 2:1 in the diet may lead to urinary calculi (water belly), which most often results in the death of the lamb.

Organic lambs are weaned at 3–4 months of age, depending on the prevailing organic regulations. Growing and finishing diets may be ground, pelleted, or consist of a mixture of whole grain and a pelleted supplement. The lambs should be adjusted to a growing diet by the time they are 2 months of age. A growing diet for lambs weighing 18 to 30kg should contain approximately 780g/kg TDN and 160g/kg crude protein. At body weights of 30kg or more the level of crude protein in the diet can be lowered to 140g/kg. Whole grain feeding combined with a protein–mineral supplement has been shown to improve the efficiency of feed conversion and rate of gain. Examples of dietary formulations as recommended by Hosford and Markus (2004) are shown in Table 5.17.

Ram Feeding

Mature breeding rams should be grazed on pasture and have a body condition score of 3.5 to 4 before the start of the breeding season. Their condition is very important at this time since they may lose up to 12% of their body weight during a 45-day breeding period. Poor nutrition is a major cause of ram mortality. In many cases, forage alone is inadequate in ensuring a proper body condition for the breeding season. Therefore thin rams should receive grain supplementation to ensure the proper condition score.

Mature rams, not being used for breeding, can be maintained on pasture or wintered on good quality hay. About 3–4kg of mixed grass and clover hay is sufficient to meet the daily energy requirements of a 115kg ram. A free choice source of water, salt, and minerals should be available at all times.

Hay for Sheep

Average or poor quality hay should be fed during gestation, leaving the higher quality hay to be fed during lactation. As protein requirements of the ewe increase dramatically after lambing, less protein supplementation from concentrate feeds is required when higher quality hay is used. Lucerne hay is an excellent feed for sheep and is best used during lactation when ewes require more protein to promote higher levels of milk production.

If lucerne hay is fed during late gestation, it should be in rationed amounts since it is very palatable and may be over-consumed. It should also be free of mould. Over-consumption of lucerne hay may result in pressure on the reproductive tract, resulting in a vaginal prolapse before lambing.

In addition, ewes receiving lucerne hay during gestation are more prone to milk fever than ewes fed grass hay. As lucerne is high in calcium, ewes are able to meet their calcium requirements without mobilizing body stores of calcium. However, after lambing, ewes not accustomed to mobilizing bone calcium may experience milk fever because of their inability to meet the additional calcium requirements associated with lactation.

Silage

High quality, finely chopped maize, grass, or small grain silage is acceptable feed for sheep. Care must be taken to properly harvest, store, and feed the silage. Poorly packed silage may contain harmful moulds, which causes listeriosis (circling disease) in sheep. Mouldy or frozen silage should be discarded and feed troughs should be cleaned daily.

Maize silage is relatively low in protein and calcium. As a result, supplementary protein, calcium, phosphorus, and vitamins can be supplied as a grain-mix top-dressed on the silage at the time of feeding.

Owing to its high moisture content, 1.5kg of silage is required to supply the energy furnished by 0.75kg of hay. The bulkiness of silage prevents adequate dry matter intake

and its use as the sole source of feed for ewes in late gestation. A typical silage-based diet fed to ewes during the last 4 weeks of pregnancy on an as-fed basis would contain about 2.5kg maize silage (35% dry matter), 1kg hay, 0.25kg maize or wheat, and 0.1kg of soya bean meal.

Minerals

It is recommended that salt (NaCl) and mineral supplementation be supplied on a free choice, year-round basis. Failure to supplement salt and minerals adequately results in low fertility, weak lambs at birth, lowered milk production, impaired immunity, and several metabolic disorders. The exact composition of these supplements should be based on knowledge of likely deficiencies in the forage available. This information can be obtained from local feed companies or advisory agencies. In general, a salt lick containing those trace minerals that are likely to be deficient (especially selenium) should be adequate during the spring and summer when sheep are grazing high quality pastures containing more than 20% clover. Complete mineral mixes are recommended when grazing low quality roughages, starting 4 weeks before breeding, during breeding, and during late gestation and early lactation.

Vitamins

Pasture or high quality hay provides all the vitamins required by most classes of sheep. However, after a drought, or when low quality hay or silage is fed, a supplement containing vitamins A, D, and E may be needed. Estimated daily vitamin requirements for ewes during late pregnancy and lactation are: 6,500 international units (IU) Vitamin A, 400 IU Vitamin D, and 40 IU Vitamin E.

FEEDING GOATS

There is an expanding market for goat meat in several countries since it is the preferred meat of certain ethnic groups. Also, there is an increasing demand for goat milk as a human food as well for the production of specialty cheeses.

The nutritional requirements of goats are similar to those of sheep, therefore a similar feeding strategy can be used (Hart, 2008; Steevens and Ricketts, 2008). A general recommendation by these researchers is to provide enough energy, protein, minerals and vitamins in a balanced feeding programme to maintain a healthy animal, and to offer does enough extra feed during gestation and lactation for foetal development and milk production.

The nutritional programme is dependent on the type of goat (meat, milk or mohair/cashmere) and the production stage. For instance, lactating dairy goats have higher nutrient requirements than dry or gestating does because of their requirements for milk production.

Energy requirements are affected by body size, growth, reproduction, and lactation, the main sources in goat diets being fibre, starch, and sugar from forages and grains. A deficiency of energy results in the loss of body condition, poor growth, reduced milking ability, and reduced reproductive performance. Likewise, the level of protein needed in the diet depends on production stage. Young, fast-growing goats need a high level of protein in the diet to deposit muscle mass, and lactating goats need a high level of protein for milk production. Inadequate protein leads to inefficient utilization of forages and reduces forage intake.

The non-pregnant, non-lactating doe has low nutritional requirements. The dry doe in good condition can be maintained on good quality pasture (or hay) with mineral–vitamin supplementation. Flushing, increasing the amount of energy fed 30 days prior to and 30 days after breeding, has been shown to increase ovulation rate in yearling does and does in poorer body condition.

Nutrient requirements increase as pregnancy progresses. During the last 6 weeks of pregnancy, the amount of supplementary grain mix or complete feed should be increased. Also, replacement does need additional feed because their bodies are still maturing.

Lactation places a great deal of stress on does, especially those nursing twins and dairy goats used for commercial milk production. Access to high quality forage or the feeding of maximum amounts of high quality hay balanced with a grain ration containing enough protein, minerals and vitamins to support production and animal health should be fed (Table 5.18). Grass or legume hays are equally acceptable. As the percentage of legumes is increased, the need for protein in the grain mix is reduced. A concentrate–grain mix or complete feed mixture should be increased gradually to promote milk production and maintain body condition. A high-producing goat will require up to 3.5kg of a complete feed daily.

To determine the amount of grain to feed, it is recommended that the level of

Forage fed	Level of protein in grain mix	Mineral mix to use
Legume or mixed, mostly legume	140 to 160g/kg	High phosphorus mix
Grass or mixed, mostly grass	160 to 180g/kg	2:1 Ca:P mix

TABLE 5.18 Example feed mixtures for a milking goat herd (from Hart, 2008; Steevens and Ricketts, 2008).

milk production, amount and quality of forages consumed, appetite and body condition be evaluated. Thin, high-producing does should have access to all the forage they can consume, plus grain to the limit of their appetite. Does in mid-lactation that are in good body condition should receive unlimited quantities of forage plus 1kg grain for each 3kg milk produced. During late lactation the does should require around 1kg grain for each 5kg milk produced.

The grain mixture can be formulated as for a lactating ewe (Table 5.19) and it is preferable that the grain be rolled or cracked grain rather than ground, to improve palatability.

An adequate supply of clean water is critical to good health and high milk production. If the water is warmed during cold weather, goats will consume more.

Feeding Dry Does

If the doe is in good body condition the amount of grain supplement can be reduced to 250–500g per day, together with all the forage she will eat. Hay fed during the dry period may be of lower quality than that fed during lactation, but the grain ration should contain additional protein. Browse, leaves and weeds are often useful to recondition the rumen. It is recommended that when the feed mix during this stage differs from the mix fed during lactation that the feed be changed to the milking forage and grain ration 2 weeks before the doe gives birth.

Kids

Kids should be allowed to nurse as soon as possible after birth in order to receive an adequate intake of colostrum. Prior to suckling, some producers wash the doe's udders and teats with warm water, then hand-milk half a cup of colostrum and feed it to the kid within 15 minutes of birth. This ensures that the newborn receives some colostrum and it provides it with protection from organisms present on the skin of the doe. Kids should then continue to nurse, and in cases of a milk shortage should be give an acceptable lamb milk replacer.

It is advisable to offer succulent forage (such as vegetable leaves or green grass) during the first week of age, to stimulate rumen development. Also, a small quantity

Feedstuff kg	Level of protein in finished mix g/kg			
	140	160	180	200
Cracked or rolled maize	380	330	270	220
Rolled oats	200	200	200	200
Soya bean meal	190	240	300	350
Beet pulp	100	100	100	100
Molasses	100	100	100	100
Trace mineralized salt	10	10	10	10
Dicalcium phosphate	18	18	18	18
Magnesium oxide	2	2	2	2
Vitamin premix*	+	+	+	+

TABLE 5.19 Example grain mixtures for lactating goats (from Hart, 2008; Steevens and Ricketts, 2008).

*Provides 2,200 units of vitamin A, 1,100 units of vitamin D and 8 units of vitamin E per kg final mixture.

of good quality hay should be provided during the first 2 weeks of age, followed by the introduction of creep feed (similar to lamb creep feed) in the third week.

Weaning can take place from about 3 months of age when the kid is eating forage, drinking water, and consuming at least 200g of starter feed daily, though the exact weaning age will be determined by the local organic regulations. Thereafter access to quality forage or a daily ration of good hay and 250g grain should allow a satisfactory growth rate.

An adequate supply of clean, fresh water should be available during this period.

Feeding Bucks

A suitable feed mixture for bucks is the same grain mix as fed to the milking herd at a rate of about 250g per day. The amount should be adjusted upwards or downwards to maintain a correct body condition and prevent them becoming fat.

Feeding Replacement Breeding Stock

As recommended by Hart (2008), replacement bucks and does must gain sufficient weight from weaning to breeding to be sufficiently large and sexually mature. Does will generally gain sufficient weight if an adequate amount of a moderate quality forage is available. If doelings are not gaining adequate weight they could be supplemented with grain such as whole shelled maize or wheat at 0.5% to 1% of body weight per day. Feeding excessive grain to doelings may result in them becoming too fat and lead to fat being deposited in the udder, which can cause reduced formation of milk secretory tissue. The doe is also more likely to have pregnancy toxemia and birthing problems.

BREEDS FOR ORGANIC PRODUCTION

Unlike the situation with pigs and poultry, which benefit from the hybrid vigour resulting from cross-breeding, pure breeds are generally preferred by organic producers. Many producers favour traditional or native breeds since they are likely to be better adapted regionally to the feed resources available on-farm, even though they may be relatively unimproved in terms of productivity.

A wide range of breeds of cattle, sheep and goats is available for organic meat and milk production internationally, displaying greatly different production and carcass characteristics and responding differently to diet composition and level of feeding. Therefore, the dietary regime and feeding programme need to be designed according to the particular breed selected.

Natural breeding is preferred in organic production rather than artificial insemination, which is used widely in conventional production. The availability of males is then an important consideration. Although artificial insemination is usually avoided by organic producers its use is justified if the alternative is a marked degree of inbreeding, which is quite common in small (particularly closed) herds.

DAIRY CATTLE

Van Diepen et al. (2007) listed the most important traits for organic dairy cattle breeding as assessed by several organic agencies and research groups (Table 5.20). The traits identified as being important in cows included high fertility, disease resistance, good conformation, strong vigour, longevity, good milking ability when forage-fed,

	Agency		
Rank of trait	Research Institute of Organic Agriculture (FiBL) Switzerland	Scottish Agricultural College (UK)	Louis Bolk Institute (LBI) Netherlands
1	Fertility	General disease resistance	Fertility
2	Cell count	Mastitis resistance	Udder health
3	Longevity	Longevity	Long productive life
4	Milk from forage	Somatic cell count (subclinical mastitis resistance)	Good milk yield/ lactation
5	Protein and fat content	Female fertility	Protein and fat content
6	Udder health	Forage intake capacity	Conformation udder
7		Feet and leg strength	Quality of legs
8		Susceptibility to lameness	
9		Resistance to parasite infestation	
10		Robustness/hardiness	

TABLE 5.20 Overview of the most important traits for organic dairy cattle breeding (Van Diepen et al., 2007).

good maternal behaviour including ease of handling, and adequate fat reserves in northern regions to provide against cold conditions.

The survey included information on the breeds of dairy and beef cattle being used on organic farms in England and Wales. The dairy breeds (and their crosses) included Jersey, Ayrshire, Maas–Rijn–Ijssel, Guernsey, British Friesian, and NZ Friesian. The beef breeds (and their crosses) included Charolais, Aberdeen–Angus, Welsh Black, South Devon, North Devon, Limousin cross-breds, Hereford, Friesian, Simmental, Galloway and Belted Galloway.

Main Characteristics of Breeds of Dairy Cattle

The outstanding characteristic of the Holstein (Friesian) is its milking ability. High yields of milk of a relatively low butterfat content are typical. The breed originated in the Netherlands. The Holstein is a large-framed animal, mature cows weighing from 550 to 650kg.

The Jersey is the second most popular breed in many parts of the world. It is the smallest of the four better-known breeds of dairy cattle, mature cows weighing between 380 and 450kg. The outstanding characteristic of the Jersey is its milk, which has a very high butterfat content. The milk (also from Guernsey cows) has a slight yellow tinge due to the presence of carotene, a precursor of vitamin A. Also, Jersey milk has the highest protein content of all the dairy breeds. Jerseys are more resistant to high environmental temperatures than Holsteins and are also better foragers.

In many respects the Guernsey has similarities to the Jersey, originating also in the Channel Islands. Ease of calving and milk with a high butterfat content are common attributes. Research has shown that 60% of Guernseys carry the Kappa Casein 'B' gene. This is of economic benefit to cheese producers, since the milk from these animals has a firmer curd, increased volume and better cheese characteristics than milk from cows without the gene.

The Guernsey is a larger animal than the Jersey. Mature cows average 450kg. Owing to the low numbers of this breed, the availability of bulls is limited.

The Ayrshire has its origins in south-west Scotland. The breed is medium-sized, mature cows weighing 450 to 500kg. Their conformation is generally considered ideal. As in the case of the Guernsey, a small population limits the availability of bulls.

The Brown Swiss is a large breed of cattle known for its docile manner, high ratio of milk protein to milk fat, and sound feet and legs. It has a good yield of meat when slaughtered. It has also a reputation for resistance to heat stress in hot and humid regions.

In deciding which breed of dairy cows is best for organic production it is useful to consider dairy farm systems in New Zealand, which are mainly pasture-based. The national dairy herd is made up of about 45% Holstein–Friesians, but that number is declining. Jerseys, Ayrshires and cross-breeds appear to be gaining popularity. Several issues have been reported in New Zealand with imported Holstein–Friesian genotypes (Verkerk, 2003). Mainly it is difficult to achieve heifer growth rates comparable with overseas rates on pasture alone. Adult stature is often less than that seen in other countries unless a significant level of supplement is included in the diet. Also, when pasture is provided as the principal or the only lactation feed, the overseas genotype experiences rapid mobilization of body reserves and an excessive loss of body condition associated with the strong drive for high milk production. Another reason is a decline in cow fertility. As in some other countries, New Zealand increased the genetic merit of cows for milk production by importation of gene-stock from other countries. The

improvement, however, appears to have resulted in a decline in cow fertility, experienced as a reduction in the proportion of cows that conceive early in the breeding season, and an overall lower survival rate.

Other researchers have addressed the issue of whether high-producing cows such as the Holstein–Friesian can produce to their full potential on pasture-based systems alone. Kolver and Muller (1998) fed high-producing dairy cows either on pasture alone or on a total mixed diet that contained concentrate. Cows fed the total mixed diet produced 40 L/cow per day and cows grazing only high quality pasture (to appetite) produced 30 L/cow per day. The lower milk yield of cows fed only on pasture was attributed mainly to a lower intake of dry matter. It was concluded, therefore, that cows of high genetic merit cannot achieve their genetic milk potential on pasture alone. The reasons for this are a greater energy expenditure in a grazed pasture system, a lower intake capacity of the cows when fed completely on a bulky feed, and an inability of the cows to maximize the utilization of pasture. Therefore, high-producing cows grazing pasture have to be supplemented with concentrates to achieve their genetic milk potential and to reduce the need to mobilize excessive amounts of body reserves in early lactation.

A review by Australian researchers (Grainger and Goddard, 2004) made the interesting observation that the Jersey cow has a larger digestive tract per unit of live weight than the Friesian or Holstein, and that this probably explains their greater intake capacity on a live weight basis. This enhanced intake capacity and ability to consume roughage could be an advantage for Jerseys in pasture-based systems.

These findings add weight to the conclusion that high-producing breeds and strains of dairy cows are not the animal of choice for the organic producer, unless the farm is able to supply substantial inputs of supplementary feed. Other breed comparisons confirm this conclusion.

For instance, Dillon et al. (2003a, b) compared the performance of Dutch Holstein–Friesian, upgraded Irish Holstein–Friesian, French Montbeliarde and French Normande dairy cow breeds in a 5-year study based on a spring-calving grass-based system of milk production. Although not an organic system, production was close to it in being forage-based. It was typical of milk production in Ireland, which is characterized by having relatively low milk production per cow and a low cost of production. The aim of the Irish system is to allow grazing to make a large contribution to the total diet of the dairy cow during lactation. Accordingly, the calving date is planned to coincide with the start of the grass-growing season. This results in a seasonal calving, pasture-based system of production, typical of organic dairy farms.

The Dutch Holstein–Friesian cows produced the highest yield of milk, fat, protein and lactose; the Normande produced the lowest, while the Irish Holstein–Friesian and Montbeliarde were intermediate. Both the Dutch Holstein–Friesian and Irish Holstein–Friesian had lower body condition scores at all stages of lactation than the Montbeliarde and Normande and the Dutch Holstein–Friesian had a significantly greater loss of body condition score over the first 8 weeks of lactation. The study showed large differences in dry matter intake between breeds, related to differences in feed requirement for milk production. For instance, a difference of 13% in grass dry matter intake was found between the Holsteins and Normandes. The results indicated that, although the Dutch Holstein–Friesian produced the most milk, much of this was at the expense of a loss of body condition in early lactation and a lower live weight gain from the middle to end of lactation.

An important finding in the Irish study was that the overall reproductive performance of the Holstein–Friesians was lower than that of the Irish Holsteins and the other breeds (Table 5.21), associated primarily with conception rate.

This lower reproductive rate was attributed to reduced fertility, namely a lower overall pregnancy rate and longer calving to conception interval. At the end of the 14 weeks after the start of breeding, significantly more of the Holstein–Friesian cows (26.3%) were not pregnant compared with the Irish

| | Breed | | | |
	Dutch Holstein-Friesian	Irish Holstein-Friesian	Montbeliarde	Normande
Calving day (day of year)	61.4	58.1	60.4	61.9
Calving to conception, days	99	87.3	82.1	82.9
Total breedings per pregnant cow	2.79	2.39	1.99	1.82
Pregnancy rate, %	73.7	83.9	91.2	91.9
Gestation length, days	284	281	288	287
Animals surviving to day 2500, %	20.6	39.7	49.2	55.8

TABLE 5.21 Effect of breed on reproductive performance of dairy cows on a seasonal grass-based system over a 5-year period (Dillon et al., 2003b)

Holsteins (16.1%) or the two other breeds (Montbeliarde 8.8% and Normande 8.1%), respectively. Similarly, the pregnancy rate to first breeding of the Holstein–Friesians was lower than in the Montbeliardes and Normandes. The Holstein–Friesian cows had more days from calving to conception than the other three breeds. Other researchers have reported similar findings, namely a negative effect of genetic selection for milk yield on reproductive performance.

Another significant finding of the Irish study was that survivability of the four breeds differed greatly. The proportion of animals that completed the 5-year production period (i.e. survived to day 2,500) was 20.6, 39.7, 49.2 and 55.8% for the Holstein–Friesians, Irish Holsteins, Montbeliardes and Normandes, respectively. The results suggest that the reproductive performance and survival of Holsteins bred for high milk production are low in a seasonal grass-based milk production system.

Nauta et al. (2009) reported a growing preference in Europe for Dutch breeds such as the Meuse Rhine Yssel breed. This dual-purpose breed already enjoys a strong position in Dutch organic dairy production. A similar breed was developed in Germany and known as the Rotbunt. The breed is known for good milk production (average 6,000 litres of milk per lactation, 4.3% milk fat and 3.5% protein in European conditions), the protein being very suitable for cheese production. These breeds are now being used in New Zealand dairy herds as first or second crosses, to improve fertility and health status as well as milk protein production. Farmers have also observed a decreased incidence of mastitis with these cross-breds. Growth rate and feed conversion compare favourably with commonly used beef breed/dairy breed comparisons, as well as carcass yield and meat quality.

Reports from Europe indicate that cross-breeding Holstein dairy cows with other breeds (Brown Swiss, Dutch Friesian, Groningen White Headed, Jersey, Meuse Rhine Yssel, Montbeliarde or Fleckvieh) reduced milk production, but improved fertility and udder health in most cross-bred animals.

BEEF CATTLE

Currently the traditional British breeds are popular among organic farmers since these breeds, mainly Aberdeen–Angus (or Angus, both Black and Red), Hereford (Horned and Polled), and Shorthorn, are well suited to grass feeding systems and produce quality meat. The Angus breed is the dominant breed in the USA, Canada, Argentina, New Zealand and Australia. The Hereford was developed originally as a dual-purpose breed.

In comparison with European breeds, the British breeds are generally smaller in mature size, reach maturity at an earlier age and have a lower growth potential. In general they have higher carcass quality. Some other indigenous breeds, e.g. Highland cattle from Scotland, are being evaluated in countries such as New Zealand because of their ability to thrive in adverse conditions and provide crofters with milk, hair and meat. One feature of breeds such as the Aberdeen–Angus, Galloway and Suffolk breeds is that they are naturally polled, a feature that is of importance to producers wishing to avoid horned breeds.

European breeds include Charolais, Chianina, Gelbvieh, Limousin, Maine Anjou, Salers, and Simmental. It is likely that these breeds evolved originally as draft animals. In comparison with British breeds, they are generally larger in mature size and reach maturity later. The carcasses are leaner than in the British breeds. The Charolais came into widespread use in the

North American cattle industry at a time when producers were seeking larger framed, heavier cattle than the traditional British breeds. They demonstrate superior growth ability and have performed well under a variety of environment conditions.

An important point in relation to the use of indigenous breeds in organic production is that their carcasses may not meet the quality standards for fat class and conformation set for conventional animals. As a result, their commercial worth may be reduced. The meat from these animals may, therefore, have to be marketed outside the usual commercial channels and be marketed instead through specialized channels.

Conventional beef cattle production involves the use of cross-breeding to confer the advantages of heterosis in cows and in slaughter stock. It is not yet clear the extent to which organic beef producers make use of cross-breeding programmes. Data cited by Blair (2011) showed that in a cross-breeding experiment involving Herefords, Angus and Shorthorns weaning weight per cow was increased by about 23% due to the effects of heterosis on survival and growth of cross-bred calves, and on reproduction rate and weaning weight of calves from cross-bred dams. More than half of this advantage was due to use of cross-bred cows. The effects of heterosis were found to be greatest for longevity and lifetime production of cows.

Many organic producers prefer to use pure-breds but it is desirable for beef animals to be cross-bred in order to obtain the full advantage of heterosis (hybrid vigour). In conventional production the cow is commonly a cross-bred (F1 generation) obtained from a crossing of selected animals of two pure breeds (e.g. Hereford × Angus). These cross-bred cows are then mated to selected bulls of a third breed such as Charolais or Limousin to impart further

heterosis and desirable carcass characteristics in the progeny (F2 generation) to be marketed as meat animals. The F2 animals do not enter the breeding herd.

Cross-breeding has been widely accepted by commercial beef producers in several countries as a method of increasing production efficiency. Approximately 70% of the conventional beef cattle marketed in the US are cross-bred. Cross-breeding is more common in beef production than in milk production.

Size of farm is no doubt a main factor determining whether or not pure breed or cross-breeds should be used, since it would require at least one bull of a second breed to be used in producing cross-bred cows, and at least one bull of a third breed to be used on the cross-bred cows, to achieve maximum heterosis.

Considerable research has been conducted to characterize and compare the major beef breeds in the USA. The most comprehensive studies have been conducted at the US Meat Animal Research Center in Clay Center, Nebraska. Since 1970, more than 30 breeds and crosses have been evaluated in a common environment and management system for various performance traits. The data provide useful comparative information on beef breeds and crosses, information that is not readily available from other sources (Greiner, 2009).

Breed group averages for birth and weaning weight, daily gain and final (slaughter) weights are shown in Table 5.22. Birth and weaning data are from both steers and heifers, whereas average daily gain and final weight are averages of steers only. Final weights were adjusted to a common age at slaughter. Significant differences among breeds for the various traits are evident. Breeds that sire calves that are heavy at birth also tend to be the heaviest at weaning, grow the fastest and have the highest final

Breed Group	% Unassisted births	% Survival to weaning	Birth wt., kg	200 day weaning wt., kg	Average daily gain, kg	Final weight, kg
Hereford × Angus	92.7	91.5	36.5	207.7	1.24	522.5
Charolais	86.8	89.5	39.2	217.3	1.31	55.3
Chianina	88.4	89.3	39.4	208.2	1.19	509.8
Gelbvieh	94.1	91.0	38.0	206.8	1.21	512.1
Limousin	91.8	90.8	36.6	200.9	1.13	489.9
Maine Anjou	79.4	88.9	39.9	206.8	1.23	520.3
Salers	95.2	91.7	36.7	210.5	1.22	520.7
Shorthorn	97.6	91.9	37.4	208.7	1.24	524.4
Simmental	89.2	88.8	38.5	207.7	1.24	520.7

TABLE 5.22 Average birth and weaning weights, daily gain and final (slaughter) weights of selected breeds of beef cattle (from Greiner, 2009).

Breed group	Carcass weight, kg	Fat thickness, cm	Rib eye area, cm2	Retail product yield %	Marbling score[1]	% USDA Choice
Hereford × Angus	320.7	1.6	72.3	67.2	543	70.7
Charolais	338.8	0.9	81.3	70.2	523	58.9
Chianina	313.9	0.8	80.0	71.9	448	27.5
Gelbvieh	311.1	1.0	77.4	70.2	507	45.2
Limousin	302.5	1.0	79.4	71.5	477	43.8
Maine Anjou	319.8	1.0	79.4	70.1	501	49.5
Salers	320.7	1.0	77.4	70.0	515	44.5
Shorthorn	320.7	1.2	71.6	67.0	566	74.7
Simmental	315.2	0.9	76.8	70.1	510	63.4

TABLE 5.23 Average carcass data of steers of selected breeds of beef cattle (from Greiner, 2009).

[1]400 = Slight degree of marbling = Select Quality Grade, 500 = Small degree of marbling = Choice Quality Grade

weights (e.g. Charolais). The high-growth breeds with heavier birth weights also tend to have more calving difficulty, resulting in a lower percentage of unassisted births. Research studies confirm that heavy birth weights are the primary cause of calving difficulty. Calf survival to weaning tends to be higher in breeds that require less assistance at birth (e.g. Hereford–Angus, Shorthorn and Salers).

Carcass data are shown in Table 5.23. Carcass weights are closely related to final weights presented above. This table demonstrates that breeds that excel in retail product yield (percentage of the carcass weight that is trimmed, saleable red meat) also have lower marbling scores and reduced percentage of USDA Choice quality grades (e.g. Chianina, Limousin). Marbling score is a measure of the amount of intramuscular fat

Weight (kg) at			
Breed	Birth	210 days	365 days
Czech Pied	33.3	234.1	375.8
Angus	29.2	241.4	379.5
B. d'Aquitaine	35.1	275.1	424.4
Hereford	24.0	195.5	308.3
Charolais	35.8	272.0	415.6
Limousin	29.2	216.0	348.2
Piemontese	37.9	207.2	341.6
Simmental	28.3	260.7	418.5

TABLE 5.24 Live weights (kg) of selected European cattle from birth to 365 days (from Jakubec et al., 2003).

in the rib eye muscle and is an indicator of eating quality. Breeds with a high marbling score generally are lower in retail product yield. Fat thickness of the carcass has the largest impact on retail product yield. As fat thickness increases, a lower percentage of the carcass is saleable retail product due to trimming loss. Consequently, lean breeds with minimal carcass fat thickness give the highest yield of retail product. Rib eye area is an indicator of total muscle mass in the carcass and has a positive influence on retail product yield. These breed differences emphasize the importance of using a combination of British and Continental genetics that complement each other in a breeding programme to produce an end product that has both acceptable carcass quality and retail product yield.

A comparison of the growth traits of eight beef cattle breeds in the Czech Republic was reported by Jakubec et al. (2003). The data are shown in Table 5.24. The comparison involved Angus, Blonde d'Aquitaine, Charolais, Czech Pied, Hereford, Limousin, Piemontese and Simmental. Live weights at birth, 210 and 365 days and average daily gains from birth to 210 days, 210 to 365 days and from birth to 365 days were recorded. The averages of Blonde d'Aquitaine were highest for all growth traits except for birth weight.

An important feature of these results was the large overall genetic variation found for growth traits, the range within breeds being between 79 and 154% of the average breed level. Producers, therefore, can expect to find considerable variation within breeds of beef cattle as well as between breeds. These differences within breeds may be as large as the differences between breeds, making selection of strain as important as selection of breed.

The creation of a National Organic Livestock Database (NOLD) in the UK is a very useful resource for organic cattle farmers (Van Diepen et al., 2007). The Soil Association's Producer Services established the database in 2001 in order to assist producers in sourcing organic replacement of specific breeds. Producers can post a request for livestock or offer livestock for sale on the database.

DUAL-PURPOSE CATTLE BREEDS

As indicated previously, dual-purpose breeds are often the animals of choice for organic milk and meat production since they meet the organic ethic more closely than breeds developed specifically as milk or meat animals. Dual-purpose breeds probably fit better with the smaller organic farms than the specialized milk or beef breeds. These are breeds such as the French Montbeliarde and French Normande. In several studies cited by Blair (2011) the Montbeliarde gave a higher profitability, due to lower replacement costs, higher beef values and acceptable milk returns.

One reservation that some organic producers have about cross-bred animals relates to the feeding programme. All of the established requirements relate to pure-breeds. Feeding programmes have, therefore, to be developed using existing information and recent research data.

An interesting analysis of the use of dual-purpose breeds was published by Abad (2010) showing that the dual-purpose or dairy–beef production system is the most widespread cattle production system adopted by small-holder farmers worldwide. Dual-purpose breeds are especially important in Latin America, where a mix of Zebu, Criollo and European breeds are used in meat and milk production. This region obtains 78% of its beef and 41% of its milk production from these breeds. In some countries of this region they produce more than 90% of the total milk produced.

Compared with specialized dairy production systems, the advantages of the dual-purpose system were: (1) reduced risk of fluctuation in milk and beef prices; (2) lower incidence of mastitis because of suckling of calves; (3) reduced need for capital investment; and (4) lower requirements of technical support. Net annual income per cow had been shown to be highest on dual-purpose farms, whether or not the cost of family labour was taken into account.

Advantages from cross-breeding can be maximized with a well-designed breeding programme that matches breeds to utilize complementarity in cows and their progeny. Near-maximum advantage can be attained by using two-breed cross dams and selected sire terminal lines. The Limousin breed often has been recommended (Fredeen et al., 1982a, b) as a terminal sire breed due to its growth characteristics and superior ability to produce lean carcasses. Limousin-sired calves also have lower birth weight and result in less calving difficulty than other Continental breeds (Vissac et al., 1982).

The main objectives of a cross-breeding policy are to maximize three traits: cow productivity, efficiency of gain in the calf (growth rate and feed conversion ratio) and suitability of the carcass for the selected market (weight, length, fat depth/quality, skin and meat and fat colour). It will be clear from this brief description of cross-breeding principles that only large organic units would be capable of producing such superior stock on a regular basis. A compromise would be to purchase bulls and cross-bred females as required, within the limits imposed by the local organic regulations.

Organic producers are encouraged to use traditional breeds that may be more suited to local conditions than improved genotypes that may have to be imported to the region. Some governments provide financial incentives for use of traditional breeds. A large number of these breeds exist in several countries, though often in relatively low numbers. One disadvantage of using a pure breed on small farms is an inadequate size of the breeding herd, leading to the problem of inbreeding that results in loss of productivity in the stock. The local climate is also an important factor influencing breed selection for outdoor-based organic production.

As stated above, beef production is frequently based on surplus animals from the dairy herd. For instance Nielsen and

Thamsborg (2002) studied dairy bull calves as a resource for organic beef production in Denmark. Some 8% of all dairy bull calves born on organic dairy farms were killed, 66% were sold to conventional farms, 6% were sold to other organic farms and 20% remained on the farm of birth. However, 59% of the farmers who sold their bull calves would have preferred to keep them. The main problems in doing so were lack of stall capacity, expected low returns and shortage of on-farm feedstuffs. The main reason for keeping the bull calves on 29% of the farms was the desire for a holistic production system. Most of these farmers had steer production (66%) because of high utilization of roughage, the capacity to graze marginal areas and their calm temperament. Marginal areas were utilized for beef production on 59% of the farms. It is concluded that the majority of the bull calves born on organic dairy farms are not reared on organic farms. This issue needs to be taken into account in any assessment of the sustainability of organic dairy farms.

SHEEP BREEDS

Sheep are multi-purpose animals, with several hundred breeds in existence worldwide. There are thought to be more sheep breeds in the UK than any other country in the world, the total population being more than 30 million and comprising more than 90 different breeds and crosses.

Generally sheep are classified according to their main product, i.e. wool, meat, milk, hides, or a combination of these. Other systems of classification include face colour (generally white or black), tail length, presence or lack of horns, and the topography for which the breed has been developed, i.e. hill/mountain or lowland breeds.

Breeds used for meat and wool production include Dorset, North County Cheviot, Suffolk, Hampshire, Oxford, Shropshire, Scottish Blackface, Texel and Rambouillet. Breeds used specially for wool include Shetland, Icelandic, Lincoln, Border Leicester and Romney. Prolific breeds, i.e. those that produce more than one lamb when giving birth, include Finnsheep and Romanov. Dolly, famous as the first mammal to be cloned from an adult somatic cell, was a Finn–Dorset cross.

Traits that are important for an organic environment include: mothering ability, hardiness, resistance to disease and parasites, and ability to forage.

Breed selection on a particular farm is based on the intended market, the type of range available and on the prevailing environmental conditions. Breeds adapted to live in rough, exposed terrain of temperate regions include the Swaledale, Scottish Blackface and Welsh Mountain. These breeds are very hardy and with a thick coat. Their lambs that are destined for market are sent to lower, more productive pastures, to be reared for meat production. Generally lambs can be moved to clean pasture at 5 to 6 weeks of age, when they begin to eat significant amounts of forage.

As with beef cattle, cross-breeding is common, to impart heterosis to the progeny. Ram (sire) breeds excel in growth and carcass (meat) characteristics whereas ewe (dam breeds) excel in fitness (e.g. longevity and parasite resistance) and reproductive traits (early puberty, prolificacy and milk production). Lambs sired by a ewe breed ram, such as Finnsheep, are usually kept as flock (ewe) replacements whereas the matings of terminal sires and cross-bred ewes are all marketed and do not enter the breeding flocks.

The most popular terminal sire breeds in North America are the Suffolk and

Hampshire. In Europe, the Texel is the most popular sire of market lambs.

GOAT BREEDS

As with sheep, there is a large number of breeds used worldwide. Generally goat production is an ancillary enterprise on the farm, rather than a primary enterprise such as cattle or pig production.

Adaptability is the most important of all the production traits. The goat is one of the most adaptable of all domesticated livestock and survives in a wide range of environments worldwide. However, when taken out of one environment and placed in another, it may not always realize its production potential. Adaptability is a lowly heritable trait, therefore will respond slowly to selection pressure.

Fertility is another important economical trait, measured by conception rate, kidding rate and the ability to breed out of season. The ability to reproduce satisfactorily is the most economically important trait when breeding animals are sold live. In general, goats have a high reproductive rate, with conception rate not being a problem. Goats that have evolved in the temperate zones of the world tend to be seasonal breeders. Females come into oestrus in the autumn with anoestrus occurring in late spring. This breeding pattern does not always coincide with the optimal marketing period of weaned kids. Goats originating in the tropics are non-seasonal breeders and breed all year round, a desirable trait.

Ability to grow well and make satisfactory weight gains is another important trait, as is health. Producers should strive to select animals that require minimal treatment for parasites and foot scald/rot.

Growth can be effectively divided into two periods: growth before weaning and growth after weaning. A high pre-weaning growth rate reflects the genetic potential of the kid but also the mothering ability of the doe.

Carcass characteristics are not regarded as important in goats as in other species such as pigs. The important carcass characteristics are dressing percentage, ratios of lean:fat:bone, and anatomical distribution of muscle. Most of the meat is in the leg and shoulder. Generally, the dressing percentage of goats is around 50%.

According to recent figures, Greece leads goat production in the European Union with more than 37.2% of the total goat population, followed by Spain (21.6%), France (9.4%), Italy (6.9%), and Romania (6.5%). About 50% of the organic herd goats in Europe are located in Greece.

Goat meat is the most widely consumed meat worldwide, mainly in Latin America, Africa, Central Asia, the Caribbean, Middle East, and parts of Europe. Meat goats are generally not used for milking, and some even have traits that make them undesirable for milk production. Most of the breeds used for meat are not used as fibre sources because they have short coats. The Spanish goat is an exception.

Meat breeds include Boer, Genemaster, Kiko, Kinder, Myotonic, Pygmy, Savanna, Spanish, Tennessee Meat Goat and TexMaster. The Myotonic goat has an unusual feature in that its muscles freeze for about 10 seconds when the goat is panicked. Though this is a painless condition and has no lasting effects, it generally results in the animal collapsing temporarily on its side.

Meat goats grow fast and are renowned brush-eaters. Some breeds are almost self-sufficient because they have evolved in feral conditions. They have the reputation of requiring minimal management because they have developed resistance to the parasites, foot rot, and respiratory problems that are common in other goats.

The Boer is a goat breed developed in South Africa. They have a slender build but are strong and vigorous. They mostly breed in autumn and early winter.

The Kiko is a breed of meat goat developed in New Zealand, Kiko being the Maori word for flesh or meat. The breed was developed from cross-bred local feral goats with imported bucks of the Anglo–Nubian, Saanen and Toggenburg breeds and has the reputation of being hardy, grows well and survives year round with minimal management.

The Spanish Meat Goat, also known as the Brush goat, was introduced to America by Europeans. They are short in stature but have a strong build. These meat goats are available in various colours and can be bred in any month of the year.

The Tennessee Fainting Goat is a Myonic goat and has a good carcass and fleece.

The Pygmy breed has an African origin, with a small body structure. Although suitable for meat production they are generally regarded as pets or hobby animals. They can be bred all year round.

Since meat goats are growing in popularity in North America, breeders are cross-breeding the existing breeds to develop better types. Some of the better-known Cross-breeds are:

Texmaster. A moderate-sized meat goat which is a cross between Boers and Myonic.

Moneymaker. Developed by first cross-breeding Saanens and Nubians and then crossing the progeny with a Boer male.

Dairy breeds include Alpine, LaMancha, Nigerian Dwarf, Nubian, Oberhasli, Saanen, Sable and Toggenburg. The Nubian, also known as the Anglo–Nubian, is considered a dual-purpose breed for milk and meat production. It was developed in England from Indian, African and European breeds of milk goats and is now the most popular breed of dairy goat in the United States.

The large Saanen goats are the most productive dairy goat breed, producing more than 3.5 litres of milk per day. This is a well-established breed, originating in Switzerland. The butterfat content of Saanen milk is lower than in some other milk breeds, at about 30g/kg.

Oberhasli, or Swiss Alpine goats, are found in several countries. These goats originated in the European Alps and are also commonly known as 'French Alpine'. These goats range from medium to large size and adapt to their environment very easily. Alpines are popular for milk production since their milk contains about 35g/kg butterfat. A disadvantage is that they are seasonal breeders.

LaMancha originated in Spain and are medium in size, with a reputation of being robust and healthy and with an excellent temperament for milking. Their milk contains about 42g/kg butterfat but they are also seasonal breeders.

Nigerian goats originated in Africa. Although they are very small they can produce up to 1.5–2 litres of milk per day. They are regarded as the best dairy goats as their milk is rich, containing about 60g/kg butterfat. They are more difficult to milk because of their size but they can bred all year round.

Nubians are large in size but are not heavy milk producers. Their milk contains 4.5% butterfat. They are considered seasonal breeders but they can be bred all year round.

Toggenburg goats are one of the oldest breed of dairy goats, with a medium-sized body. They are commonly raised for producing milk used to make cheese. They have long lactations, their milk containing about 33g/kg of butterfat.

Kinders are dual-purpose goats, being developed in the USA for both milk and meat. An advantage is that they are not seasonal breeders and can be bred at any time of the year.

The main fibre breeds are the Angora, Cashmere and Pygora.

Angoras are raised for their thick fleece. They are medium sized goats having a long thick coat also known as mohair. They originated in Turkey.

Pygora is a cross-breed of Pygmy and Angora raised to produce fine fibre. They can be shorn twice a year.

HEALTH

Dairy Cattle

As reported by Marley et al. (2010), the incidence of clinical and sub-clinical diseases recorded in conventional dairy systems is similar to that in organic dairy systems, with infertility, lameness and mastitis being the major problems. However, the severity of these diseases may be either lower or higher in organic systems due to different management practices.

Some studies have shown a lower incidence of mastitis in organic dairy herds, which is usually explained by a lower milk production in cows fed organically.

The overriding importance of management was demonstrated in an American study in which the health of cows on organic and conventional dairy farms in three regions of the USA was compared. Milk samples were screened for bacteria and common diseases, also records on veterinary help were maintained. The 5-year project investigated several aspects of dairy cow health, including nutrition, lameness and udder cleanliness. The results showed no significant differences in health or in the nutritional content of the milk (Bergman, et al., 2014). A total of 192 organic and 100 conventional dairy farms in New York,

Oregon and Wisconsin participated in the study.

The study found also that more conventional farms (69%) used veterinarians than organic farms (36%). Fewer hock lesions were reported on organic farms. Some organic herds in the study were found to be carrying a strain of bacteria (*Streptococcus agalactiae*) no longer found in conventional herds due to the prior use of antibiotics.

Few farms in this study performed well in formal criteria used to measure the health and well-being of cows, suggesting that management on both types of farms needs to be improved:

- Only one in five herds met standards for hygiene, a measure of animal cleanliness
- 30% of herds met proper criteria for body condition score
- 26% of organic and 18% of conventional farms met recommendations for pain relief during dehorning
- 4% of farms fed calves the recommended amount of colostrum
- 88% of farms did not have an integrated plan to control mastitis
- 42% of conventional farms met standards for treating lameness
- Cows on the organic farms produced 25% less milk per day than conventional grazing herds.

Good nutrition can be effective in maintaining an optimal immune response and defence against attacks from pathogens but, as reported by Kijlstra and Eijck (2006), the effect of organic feeding on immunological defence in organic livestock has not been reported. Based on other findings, nutrients such as glutamine, arginine, certain fatty acids, vitamins E and C, zinc and selenium would be expected to influence the immune response. A related aspect in grazing animals is that poor nutrition may predispose the animal to intestinal worm infestation. As suggested by Kijlstra and Eijck (2006), organic feed may be deficient in some important nutrients and research on nutrition in relation to the immune response in organic livestock production is required.

Other reports confirm that herd health problems on organic dairy farms are similar to those found on conventional dairy farms. The main problems have been identified as: mastitis, fertility disorders and hoof diseases, with metabolic disorders such as acetonaemia and milk fever being less frequent (Krutzinna et al., 1996). No differences in the incidence of mastitis or somatic cell counts were reported in a Danish study involving 27 organic and 57 conventional herds (Vaarst and Bennedsgaard, 2001).

Beef Cattle

No specific health problems have been reported in organic beef cattle (Nielsen and Thamsborg, 2002). Nutritional deficiencies in forage are more likely to occur in beef cattle than dairy cattle, since beef animals spend more time on range than dairy cattle and dairy animals are more subject to frequent inspection. Nutritional problems that may be encountered in beef animals include the following (Merck Veterinary Manual, 2010):

Ataxia affects calves mainly and is most often attributed to a chronic manganese deficiency. Deformities of affected animals include weak legs and pasterns, enlarged joints, stiffness, twisted legs, general weakness and reduced bone strength. It can also be caused by a potassium deficiency.

'**Blind staggers**' is a sign of acute selenium toxicity. Affected cattle show dullness, ataxia, rapid weak pulse, laboured respiration, diarrhoea and lethargy; the head is lowered and the ears droop. Death is due to respiratory failure.

Bloat or tympanites of the rumen occurs when the gases produced during fermentation cannot be expelled through eructation and cause inflation and swelling of the rumen. Bloat can be found in cattle grazing lush, young legumes. It is advised, therefore, that animals be introduced gradually to pastures containing legumes. The condition may require veterinary attention.

Cardiac arrhythmia is usually associated with a prolonged and severe deficiency of sodium in the diet.

Delayed puberty is largely attributable to an inadequate content of energy in the diet fed to young, growing animals.

Depraved appetite (pica) is seen when cattle consume non-feed materials such as soil, sand or fine stone, or engage in persistent licking, chewing or eating of wood and many other substances for no apparent reason. Many suggest that these habits can be explained on the basis of nutrient deficiency, but this has not been confirmed by research. Sporadic cases may indicate a brain disorder or poisoning from ingestion of a weed such as ragwort. More extensive outbreaks may be due to parasitism or mineral deficiency and should be investigated by blood and feed tests.

Dermatitis can be seen in calves and older cattle due to a zinc deficiency. Generally, it is most severe on the legs, neck and head and around the nostrils. Wounds are slow to heal. Additional signs associated with zinc deficiency include decreased testicular growth, listlessness, development of swollen feet with open scaly lesions and alopecia (hair loss).

Dystrophic tongue, in which the tongue surface is degenerated, is most common of the overall selenium deficiency syndromes. It can also be caused by a deficiency of vitamin E.

Goitre (thyroid gland enlargement) is a sign of an iodine deficiency. Affected cows may give birth to hairless calves. The condition may occur in cattle consuming diets with an adequate level of iodine when fed crops of the cruciferae family, such as turnips or cabbage. These crops may contain a goitrogen which interferes with iodine uptake by the thyroid gland. The cyanogenetic goitrogens include a thiocyanate found in white clover and glucosinolates found in some brassica forages such as kale, turnips, and rape. They impair iodine uptake by the thyroid gland, and their effect can be overcome by increasing the dietary iodine content.

Hair coat roughness may be related to deficiency of energy, phosphorus, salt, vitamin A, cobalt, or copper.

Heart failure is often associated with a selenium deficiency.

Haemoglobinemia most often is a manifestation of a copper deficiency.

Hypomagnesemic tetany (grass tetany) is due to a relative deficiency of magnesium – relative in that the dietary magnesium may be bound in such a manner that it is not bio-available. Among the signs of experimentally produced magnesium deficiency in both young and mature cattle are anorexia, hyperemia, greatly increased excitability and calcification of soft tissues in a chronic deficiency condition. An affected animal exhibits convulsions, falling on its side with its legs alternately extended and relaxed. Death may occur during the convulsions. Frothing at the mouth and profuse salivation are evident. The signs appear to progress much more rapidly in adult cows. Animals showing clinical signs require treatment immediately with combined solutions of calcium and magnesium. The problem can be prevented by ensuring an adequacy of bio-available magnesium in the diet.

Mastitis is inflammation of the mammary gland. Mastitis may possibly be alleviated by supplementation with β-carotene. This disease and resulting infection can significantly reduce milk production. Mastitis is most commonly found in dairy herds, but it can also occur in beef herds, resulting in a reduction in weaning weight.

Polioencephalomalacia is characterized by listlessness, muscular incoordination, progressive blindness, convulsions and death. It is linked to some aspect of the diet that produces high levels of thiaminase, which destroys thiamine (one of the B vitamins). Affected cattle are very responsive to treatment with thiamine, preferably via IM injection. Thiamine is involved in the normal functioning of the central nervous system as well as other systems.

Retained placenta. The nutritional causes of retained placenta appear to be rather complex and include deficiencies of selenium, vitamin A, copper, and iodine. Prepartum injection of selenium has been shown to reduce the incidence of retained placenta.

Many or all of the above problems can be prevented by the use of salt licks and mineral blocks formulated to correct any deficiencies likely on the pasture and range areas in question.

Sheep

Vaccination against certain diseases may be permitted in organic production and should be used to the maximum extent permitted by the local organic certification regulations. The most common diseases that can be prevented by vaccination are Clostridial diseases such as enterotoxemia caused by *Clostridium perfringens* types C and D and tetanus caused by *Clostridium tetani*.

Under the Canadian regulations vaccinations are allowed for diseases that cannot be controlled by other management techniques. Treatments must be recorded and an increased withdrawal time, usually twice that recommended by the manufacturer, is required.

It is clear that the most difficult aspect of raising organic lamb is controlling internal parasites. Sheep are particularly susceptible because they swallow parasites as they graze close to the ground. Regular use of chemical de-wormers to control this problem is not allowed in organic systems, although a restricted use of medication is allowed in some countries such as the UK. Where chemical de-worming is allowed, treatment of mature ewes is advised in order to prevent infection of young animals. The recommendation is to use them one month prior to lambing, after lambing before going on to clean pasture, and/or before moving ewes to new pasture.

Organic certification agencies vary in their approach to the use of de-worming products and prevention strategies are encouraged as an alternative. Certification requirements associated with the use of medication vary, therefore it is advisable to consult with the local certification body before using any de-wormer or medication. Various preparations have been suggested as an alternative to chemical de-wormers. However, research conducted in British Columbia (Canada) tested a herbal preparation, garlic, diatomaceous earth and pyrethrum and failed to show that any was effective.

Management strategies recommended for the control of internal parasites include the use of grazing programmes designed to minimize exposure to parasites, providing good nutrition and minimizing stress. Some research has shown that sheep provided with high-protein diets are more resistant to parasites, also inclusion in the diet of forage high in tannins such as birdsfoot trefoil is partly effective. It is advisable to move lambs to clean pasture at weaning time and to use forward creep grazing to provide lambs with clean pastures before the ewes. Grazing young lambs on contaminated pasture should be avoided, also grazing lambs on the same permanent pasture 2 years in a row.

Intestinal worms result in the loss of large quantities of blood and protein, causing a lack of condition, weakness and anaemia. The wool may become brittle and fall out. Anaemia is characterized by paleness of the gums and the linings of the eyelids, providing a way of assessing the animals.

Sheep are particularly susceptible to these parasites and unfortunately sheep are unable to develop a fully-effective immunity. One reason for their susceptibility to internal parasites is that the small faecal pellets of sheep disintegrate very easily, thus releasing the worm larvae on to pastures once the animals are infected. The feeding behaviour of sheep that causes them to graze close to the ground where numbers of larvae are higher increases their exposure to parasites. Their flocking instinct that encourages them to graze close together and their tendency to graze areas of high faecal contamination are additional factors. The parasitic larvae develop and survive best under warm, wet conditions and they may survive on pastures through the winter. Some species of intestinal worms produce eggs in large numbers, so that worm numbers can build up very rapidly.

Coccidiosis is another condition that may affect young lambs. It is caused by single-celled parasites (protozoa) called Eimeria, which undergo a simple life cycle in the gut and are shed in the manure. This parasite can survive on the ground for up to a year. Only some lambs appear to develop clinical signs, with animals showing symptoms such as lethargy, lack of appetite, diarrhoea, dehydration, weight loss and anaemia. Preventative measures such as ensuring a low stocking density and outdoor lambing management systems are advised, in order to reduce the risk of clinical coccidiosis. Treatment with chemical agents is allowed under some organic certification systems.

Goats

As with sheep, vaccination against certain diseases may be permitted in organic production and should be used to the maximum extent permitted by the local organic certification regulations. For instance, goats are susceptible to enterotoxemia (over-eating disease), caused by Clostridial organisms (*Clostridium perfringens* type C and D) that are present in the digestive tract. Under normal conditions, these potential pathogens do not cause harm. However, stress (environmental, physiological or psychological) may trigger the disease, which is often fatal in kids. Does can be vaccinated for CDT at about 30 days prior to giving birth, in order to provide protection to the newborns via colostrum. The kids are then vaccinated at 5–6 weeks of age, followed by a booster shot 3–4 weeks later.

As with sheep, internal parasites are difficult to control in organic goats. Organic standards in Europe and the UK generally allow de-worming of small ruminants because they recognize that totally natural control of internal parasites is difficult to achieve and generally compromises the welfare of lambs and goats. For organic producers, a practical approach is to minimize the use of chemical treatments until it can be demonstrated that it is possible to eliminate chemical treatments entirely.

Good nutrition is the other main approach to ensuring the optimal health of goats and minimizing the potential for them to be affected by common problems such as parasites. Lu et al. (2010) outlined the importance of nutrition in controlling parasites in goats, showing that protein supplementation can improve resistance and enhance the production efficiency of organic herds. An adequate supply of energy and protein is also important in achieving optimal microbial synthesis in the rumen and efficient production in grazing animals.

Grazing management strategies are also important in the control of internal parasites and diseases. These include keeping the stocking rates low, ensuring that parasite-free animals are turned out on to clean pastures and rotating from contaminated to clean pastures. Some goats can develop a measure of immunity to parasites and should be selected as potential breeding stock.

As the use of medication is restricted in organic goat production, an active area of research is the search for non-chemical treatments for the prevention and control and treatment of parasitic infections in goats (Moreno-Gonzalo et al., 2012). Some plants and plant preparations have been identified as alternative de-worming agents. The list includes black cumin, black walnut, boundary tree, common mugwort, common wormwood, crucifers, custard tree, eucalyptus, Eurasian wormwood, fargara, fennel, fern, fumitory garlic, Gambian mahogany, goosefoot, Indian lilac, kamala tree, neem tree, papaya, pinkroot, pumpkin, pyrethrum, sacred basil, southern wormwood, tansy, tarragon, wild carrot, and wild ginger. A number of legumes (sainfoin, sulla, *Lotus corniculatus*, *Lotus pedunculatus*, dorycnium, chicory) and woody plants (heather, oak, hazel tree, blackberries) have also been identified as being useful in parasite and disease prevention, control and treatment in organic goat production.

European research is showing promising results with the use of heather supplementation to control certain types of intestinal worm infestations in goats and improve production (Moreno-Gonzalo et al., 2012). The researchers concluded that goats supplemented with heather had a greater resilience to worm infestations in the gut and that heather significantly reduced the number of worm eggs secreted by grazing goats. The factor responsible in heather is likely to be the tannin content.

Producers planning to use any of these treatments should do so following veterinary advice.

REFERENCES

Abad, R. (2010) Dairy-beef or dual purpose cattle production. Available at: http://www.agri-pro-focus.nl (accessed April 20, 2010).

ADAS (2005) Final Report on Project BD1415 to DEFRA, EN, CCW and DARDNI. Role of Organic Fertilizers in the Sustainable Management of Semi-natural Grasslands. ADAS, Wolverhampton, UK, 13 pp.

AFRC (1993) Energy and Protein Requirements of Ruminants. CAB International, Wallingford, UK.

Bergman, M.A., Richert, R.M., Cicconi-Hogan, K.M., Gamroth, M.J., Schukken, Y.H., Stiglbauer, K.E. and Ruegg, P.L. (2014). Comparison of selected animal observations and management practices used to assess welfare of calves and adult dairy cows on organic and conventional dairy farms. Journal of Dairy Science 97:4269-4280.

Blair, R. (2007) Nutrition and Feeding of Organic Pigs. CAB International, Wallingford, UK, 322 pp.

Blair, R. (2008) Nutrition and Feeding of Organic Poultry. CAB International, Wallingford, UK, 314 pp.

Blair, R. (2011) Nutrition and Feeding of Organic Cattle. CAB International, Wallingford, UK, 293 pp.

Blair, R. (2011) Organic Production and Food Quality: A Down to Earth Analysis, Wiley-Blackwell, Oxford, UK, 296 pp.

Brigstocke, T.D.A., Cuthbert, N.H., Thickett, W.S., Lindeman, M.A. and Wilson, P.N. (1981) A comparison of a dairy cow compound feed with and without cassava given with grass silage. Animal Production 33: 19–24.

Cherney, D.J.R., Cherney, J.H. and Chase, L.E. (2009) Using forages in dairy rations: are we moving forward? Proceedings of the Cornell Nutrition Conference, pp. 202–209.

Chiba, L.J. (2009) Animal Nutrition Handbook, Section 15: Dairy cattle nutrition and feeding. Available at: http://www.ag.auburn.edu/~chibale/an15dairycattlefeeding.pdf (accessed on: November 22, 2009).

Corbett, J. L. (1980) Grazing ruminants: evaluation of their feeds and needs. Proceedings of the New Zealand Society of Animal Production 40: 136–144.

CSIRO (2007) Nutrient Requirements of Domesticated Ruminants. CSIRO Publishing, Collingwood, Victoria, 296 pp.

Dillon, P. (2009) Grazing Notebook. Teagasc Dairy production research Centre, Moorpark, Fermoy, Cork, Ireland.

Dillon, P.G., Buckley, F., O'Connor, P., Hegarty, D. and Rath, M. (2003a) A comparison of different dairy cow breeds on a seasonal grass-based system of milk production 1. Milk production, live weight, body condition score and DM intake. Livestock Production Science 83: 21–33.

Dillon, P.G., Snijders, S., Buckley, F., Harris, B., O'Connor, P. and Mee, J.F. (2003b) A comparison of different dairy cow breeds on a seasonal grass-based system of milk production. 2. Reproduction and survival. Livestock Production Science 83: 35–42.

Fredeen, H.T., Weiss, G.M., Lawson, J.E., Newman, J.A. and Rahnefeld, G.W. (1982a) Environmental and genetic effects on preweaning performance of calves from first-cross cows. 1. Calving ease and preweaning mortality. Canadian Journal of Animal Science 62: 35–49.

Fredeen, H.T., Weiss, G.M., Rahnefeld, G.W., Lawson, J.E. and Newman, J.A. (1982b) Environmental and genetic effects on preweaning performance of calves from first-cross cows. II. Growth traits. Canadian Journal of Animal Science 62: 51–67.

Grainger, C, and Goddard, M.E. (2004) A review of the effects of dairy breed on feed conversion efficiency – An opportunity lost? Animal Production in Australia 25, 77–80.

Greiner, S.P. (2009) Beef Cattle Breeds and Biological Types. Virginia Co-operative Extension Publication number 400–803, Virginia Polytechnic Institute and State University, Blacksburg, USA.

Haas, G., Deittert, C. and Köpkea, U. (2007) Farm-gate nutrient balance assessment of organic dairy farms at different intensity levels in Germany. Renewable Agriculture and Food Systems 22: 223–232.

Hart, S. (2008) Meat Goat Nutrition. Pages 58–83 in Proc. 23rd Ann. Goat Field Day, Langston University, Langston, OK, USA.

Hosford, S. and Markus, S. (2004) Feeding Lambs – Frequently Asked Questions. Alberta Agriculture and Rural Development, Alberta Agriculture and Forestry, Edmonton, Alberta.

INRA (1989) Ruminant nutrition: recommended allowances and feed tables (Jarrige R, éditeur). INRA Paris, 389 pp.

Jakubec, V, Schlote, W., Riha, J. and Majzlik, I. (2003) A comparison of growth traits of eight beef cattle breeds in the Czech Republic. Archiv für Tierzucht 46, 143–153.

Kersbergen, R. (2010a) Maximizing Organic Milk Production and Profitability with Quality Forages. University of Maine Co-operative Extension. Available at: http://www.extension. org/article/24980, accessed May 12, 2010.

Kersbergen, R. (2010b) personal communication.

Kijlstra, A. and Eijck, I.A.J.M. (2006) Animal health in organic livestock production systems: a review. NJAS Wageningen Journal of Life Sciences 54: 77–94.

Kolver, E.S. and Muller, L.D. (1998) Performance and Nutrient Intake of High Producing Holstein Cows Consuming Pasture or a Total Mixed Ration. Journal of Dairy Science 81: 1403–1411.

Krutzinna, C., Boehncke, E. and Herrmann, H.J. (1996) Organic milk production in Germany. Biological Agriculture and Horticulture 13: 351–358.

Lu, C.D., Gangyi, X. and Kawas, J.R. (2010) Organic goat production, processing and marketing: Opportunities, challenges and outlook. Small Ruminant Research 89: 102–109.

Marley, C.L., Weller, R.F., Neale, M., Main, D.C., Roderick, S. and Keatinge, R. (2010). Aligning health and welfare principles and practice in organic dairy systems: a review Animal 4: 259–271.

Merck Veterinary Manual (2010) Nutritional Diseases of Cattle. http://www.merckvetman ual.com/mvm/index.jsp?cfile=htm/bc/182315. htm, accessed May 5, 2010.

Moreno-Gonzalo, J., Ferre, I., Celaya, R., Frutos, P., Ferreira, L.M.M., Hervás, G., García, U., Ortega-Mora, L.M. and Osoro, K. (2012) Potential use of heather to control gastrointestinal nematodes in goats. Small Ruminant Research 103: 60–68.

NAS–NRC (1985) Nutrient Requirements of Sheep, 6th Revised edn. National Research Council, National Academy of Sciences, Washington, DC.

NAS–NRC (1989) Nutrient Requirements of Dairy Cattle, 7th Revised edn. National Research Council, National Academy of Sciences, Washington, DC.

NAS–NRC (1994) Nutrient Requirements of Poultry, 9th Revised edn. National Research Council, National Academy of Sciences, Washington, DC.

NAS–NRC (2000) Nutrient Requirements of Beef Cattle, 7th Revised edn. National Research

Council, National Academy of Sciences, Washington, DC.

NAS–NRC (2001) Nutrient Requirements of Dairy Cattle, 8th Revised edn. National Research Council, National Academy of Sciences, Washington, DC.

NAS–NRC (2007) Nutrient Requirements of Small Ruminants: Sheep, Goats, Cervids, and New World Camelids, Revised edn. National Research Council, National Academy of Sciences, Washington, DC.

NAS–NRC (2012) Nutrient Requirements of Swine, 11th revised edn. National Research Council. National Academy of Sciences, Washington, DC.

Nauta, W., Baars, T., Saatkamp, H., Weenink, D. and Roep, D. (2009) Farming strategies in organic dairy farming: Effects on breeding goal and choice of breed. An explorative study. Livestock Science 121: 187–199.

Nielsen, B. and Thamsborg, S.M. (2002) Dairy bull calves as a resource for organic beef production: a farm survey in Denmark. Livestock Production Science 75: 245–255.

Steevens, B. and Ricketts, R. (2008) Feeding, Managing and Housing Dairy Goats. Department of Animal Sciences, University of Missouri-Columbia, Agricultural publication G3990, http://muextension.missouri.edu/xplor/agguides/dairy/g03990.htm, accessed January 5, 2016.

Stockdale, C.R. (1999) Effects of cereal grain, lupins, cereal grain or hay supplements on the intake and performance of grazing dairy cows. Australian Journal of Experimental Agriculture 39: 811–817.

Umberger, S.H. (2016) Feeding Sheep. Virginia Co-operative Extension Publication 410–853, Virginia Polytechnic Institute and State University, Virginia State University, Blacksburg.

Vaarst, M. and Bennedsgaard, T.W. (2001) Reduced medication in organic farming with emphasis on organic dairy production. Acta Vet Scand Suppl. 95: 51–57.

Van Diepen, P., McLean, B. and Frost, D. (2007) Livestock Breeds and Organic Farming Systems, Farming Connect Report for Organic Centre Wales. Available at: http://orgprints.org/10822/1/breeds07/pdf, accessed April April 2010.

Verkerk, G. and Tervit, R. (2003) Pasture based dairying: challenges and rewards for New Zealand producers. Theriogenology 59: 553–561.

Vissac, B., Foulley, J.L. and Menissier, F. (1982) Using breed resources of continental beef cattle: the French situation. In: Barton, R.A. and Smith, W.C. (eds) Proceedings of the World Congress of Sheep and Beef Cattle Breeding. Dunmore Press, Wellington, New Zealand, pp. 101–113.

Weary, D.M. (2001) Calf management: improving calf welfare and production. Advances in Dairy Technology 13: 107–118.

Weller, R.F. (2002) A comparison of two systems of organic milk production. In: Kyriazakis, I. and Zervas, G. (eds) Proceedings Organic Meat and Milk from Ruminants, Athens, 4–6 October 2002, EAAP Publication 106, pp. 111–116.

Younie, D. (2001) Organic and conventional beef production – a European perspective. In: Proceedings of the 22nd Western Nutrition Conference, Saskatoon, Canada.

Younie, D. and Mackie, C.K. (1996) Factors affecting profitability of organic, low-input and high-input beef systems. In: Parente, G., Frame, J. and Orsi, S. (eds) Grassland and Land Use Systems, 16th Meeting of European Grassland Federation, Grado, Italy, September 1996, pp. 879–882.

CHAPTER 6

Integrated Systems

A main objective of organic livestock farming is to maximize the resources of the farm in a sustainable and effective manner and in a way that is as natural as possible. This can be achieved mainly by multi-species grazing, i.e. the practice of using two or more species of livestock together or separately on the same land in a specific growing season.

Different species of livestock prefer different forages and graze them to different heights. With an understanding of the different grazing behaviours of each species, various combinations of animals can be used to more efficiently utilize the forages in a pasture. Grazing cattle, sheep and goats together on a diverse pasture should result in all types of plant material available being consumed, resulting in a more efficient utilization of the forage and browse.

However, more research needs to be done in this area so that organic producers have more specific guidelines for implementation on any particular farm. Currently it is necessary for producers to test out several systems until the most satisfactory one is

FIGURE 6.1 Mixed grazing.

identified. In general, current research findings (Coffey, 2001; Pennington, 2014) indicate that multi-species grazing can yield a more efficient and uniform use of pastures, but that results will vary with the type of pasture, land type and climatic conditions. Land that includes grasses, forbs and browse are best utilized with multi-species grazing. If the terrain is steep and rough, goats and sheep make better use of the available grazing land than cattle. They also eat more forbs and browse than cattle, since sheep and goats are well adapted to grazing rough borders around an otherwise relatively level pasture.

Varying terrain also lends itself to multi-species grazing. Cattle prefer to graze grass and prefer more gently sloping land. Land that is uniformly in grass may best be utilized for cattle.

Pigs do not pasture well, in that they root in the soil and plough up the ground, making it unsuitable for grazing unless nose rings have been inserted. Also, sows with piglets at foot are liable to attack anything they consider a threat. Boars tend to be aggressive and are easier to manage when penned separately. Wooded areas that can be fenced off are well suited to outdoor pig production.

Poultry can be added to the mix of animals on pasture, making a much lower usage of plant material than ruminant animals and ingesting seeds, earthworms and insects, etc., not consumed by other stock. The availability of a lake or pond on the farm would suggest the addition of waterfowl to the mix of species, as well as fish to utilize the resources presented by the water. Geese can also serve as 'watchdogs'.

Multi-species grazing can improve utilization of forages by around 5–20%, depending primarily on the type of vegetation, land type and the mix of animals used. It is the combination of grasses, forbs and browse

that provides for the more efficient use of multiple species for grazing, sometimes increasing meat production per hectare by more than 20%.

As explained by Pennington (2014), cattle tend to be intermediate grazers. They graze grasses and legumes and bite with their mouth and tongue. Cattle and horses tend to graze grasses better than small ruminants such as sheep and goats. Sheep and goats consume forbs (many of which are weeds) better than cattle or horses, although goats have a greater preference for browse than sheep.

It has been shown that sheep graze near cattle manure deposits, which cattle avoid. This results in more even use of the pasture, also in higher carrying capacity and pasture productivity.

Improved brush and weed control is one noticeable benefit from multi-species grazing with cattle and small ruminants. Sheep and goats can be used to consume weeds and browse that cattle avoid. Some of these weeds are problematic in certain areas. For example, leafy spurge and larkspur if ingested by cattle are harmful, but can be consumed safely by sheep. Using sheep to control these weeds results in safer pastures for cattle and an overall better utilization of the available pasture.

The addition of goats to cattle pastures has been shown to benefit the cattle by reducing browse plants and broad-leaved weeds. This allows for better grass growth. Goats will control blackberry brambles, multiflora rose, honeysuckle and many other troublesome plants. This is a simple and cost-effective way of renovating pastures. The same principle holds for sheep. Although they are less likely to clean up woody plants, sheep are quite effective at controlling several weeds.

Another benefit of a multi-species grazing system is the effective control of internal parasites in sheep and goats. Worm infestations are a major concern with sheep and goats, especially under organic conditions that restrict or prohibit the use of chemical treatments. Worm eggs from affected animals are deposited on the pasture in the manure and the eggs hatch and larvae are consumed by grazing animals, resulting in reinfection and the cycle of infestation being repeated. If left untreated, the concentrations of parasites will increase. These parasites are mostly species-specific, i.e. cattle parasites affect cattle, and not sheep, while sheep parasites affect sheep but not cattle. Thus, cattle can be used on affected pastures, ingesting the sheep worm larvae, and preventing them from affecting the sheep. This is most helpful when sheep and cattle follow each other in a grazing system. However, goats and sheep do share parasites, and therefore grazing them together does not improve parasite control.

As parasite eggs are deposited in the manure, and larvae only travel a short distance up grass blades, animals grazing taller forages (well above ground level) will not consume worm eggs or larvae. Therefore, goats that are given ample browse will be much less likely to become infested with parasites. If goats are forced to graze at ground level, however, the goats may acquire a serious parasite load.

Since gastrointestinal parasites affecting sheep and goats do not survive in the gut of cattle (and vice versa), it is recommended that fields infected with a high concentration of larvae from sheep and goat parasites should be grazed first with cattle to remove as many of the larvae of parasites as possible, so that sheep and goats can then graze with less danger of parasite infestation. Producers may wish to seek veterinary advice on this important issue.

The identification of effective medication for parasitic control that is acceptable to the organic industry continues to be an urgent area of research. A related area is the breeding of sheep and goat types with increased resistance to parasites.

Wildlife existing in the area of the farm with several species of livestock may be carriers of several pathogens, which can be transmitted to the stock and possibly the farm staff. Veterinary advice may have to be obtained on this issue.

A potential problem with grazing of multiple species is the feeding of supplemental trace minerals. The mineral supplement that is adequate in copper for sheep is likely to be inadequate for cattle, and a mineral supplement that is best for cattle may be toxic to sheep. Therefore precautions should be taken to provide separate mineral supplements to sheep and cattle.

PERSPECTIVE ON FUTURE DEVELOPMENTS
Environmental aspects

A current concern relating to animal production is its contribution to greenhouse gas production, particularly methane. This gas is considered to have 21 times the global warming potential of carbon dioxide. It has been estimated that beef production worldwide accounts for about 62% of total livestock methane emissions, milk 19%, sheep 12%, pigs 5% and poultry 1%. Estimates suggest that livestock in Asia and the Pacific produce 33% of total methane emissions, Latin America 23%, Europe 14%, Africa 14%, North America 11% and Oceania 5% (Blair, 2011).

Research findings indicate that organic cattle production is not as benign in terms of greenhouse gas production as it might appear initially (Blair, 2011). As explained above in the section on Digestion of

Carbohydrates by dairy cattle, fibrous diets promote a higher production of methane than diets that are more readily digested. This explains why organic milk production inherently increases methane emission, as reported by DeBoer (2003), unless the animals are fed highly digestible diets. Similarly, organic beef cattle emit more methane than conventional beef cattle.

As outlined in Chapter 5, a common perception is that traditional pasture-based, low-input dairy systems are more in keeping with environmental stewardship than modern milk production systems. This concept was examined by Capper et al. (2009), who compared the environmental impact of US dairy production in 1944 and 2007. They calculated that the carbon 'footprint' per billion kg of milk produced in 2007 was 37% of that related to the same amount of milk produced in 1944.

Farming methods have also been compared internationally from an environmental aspect. The situation was shown dramatically in a comparison of a dairy farm in Wisconsin with one in New Zealand (Johnson et al., 2002). Using total farm emissions per kg milk produced as a parameter, the researchers showed that production of methane from belching was higher in the New Zealand farm, while carbon dioxide production was higher in the Wisconsin farm. Output of nitrous oxide, a gas with an estimated global warming potential 310 times that of carbon dioxide, was also higher in the New Zealand farm. Methane resulting from manure handling was similar on the two types of farm.

The environmental impact of animal farming (conventional and organic) is therefore an issue that is currently being reviewed by scientists. One response to the issue is that the European Common Agricultural Policy has been revised to add supplementary measures that include the environmental role of agriculture. One measure is the adoption of a life cycle assessment to estimate emissions per kg of carbon dioxide equivalent per kg of live weight leaving the farm gate per annum and per hectare. It is possible that in some countries the adoption of legislation of this type will place restrictions on ruminant animal production, including organic farms.

An obvious response on organic farms to address the issue and minimize the output of greenhouse gases is to ensure that pastures are improved as much as possible.

Researchers have addressed this important issue. As explained in Chapter 5, as the pattern of ruminal fermentation in cattle, sheep and goats alters from acetate to mainly propionate, both hydrogen and methane production are reduced. This relationship between methane production and the ratio of the various VFAs is well documented. It explains why the feeding of very fibrous diets results in more methane than less fibrous diets. The fibrous diets promote a higher production of acetate in the rumen, resulting in more hydrogen and more methane. Methane is formed as a result of the need to remove hydrogen from the rumen. Diets that are more highly digestible promote a higher production of propionate, which results in less methane in the rumen. There is also an economic aspect to methane reduction in cattle. Methane represents a significant loss of dietary energy, thus reducing methane production in the gut may also improve feed efficiency. This aspect is therefore being researched actively and several recommendations have been made, some of which are relevant to organic production methods.

According to Chase (2008) and other researchers such as O'Mara et al. (2008), there are a large number of potential approaches that could be used to decrease total methane emissions from dairy cattle

Milk kg/day	DM intake kg/day	Methane produced L/day	Methane produced L/kg milk
20	16.8	518	26.0
30	19.5	580	19.4
40	23.6	652	16.3
50	28.2	725	14.5
60	33.2	793	13.2

TABLE 6.1 Relationship between level of milk production and methane emissions in dairy cattle (Chase, 2008).

and to lower the methane produced per kg of milk. The main approaches that could be undertaken by organic farmers include the following.

1. **Improved animal productivity.** The information in Table 6.1 shows the relationship between daily milk production and methane emissions in dairy cattle fed the same diet. It can be seen that as milk production increases the methane produced per cow per day increases. This is logical since the animal is consuming and processing larger quantities of feed to produce an increased yield of milk. However, the amount of methane generated per unit of milk produced decreases as milk production increases. The net effect is that fewer animals are needed to produce a specific quantity of milk, resulting in less total methane being generated. Other factors related to increasing animal productivity include genetics, feed quality, ration formulation and animal management.

2. **Use of high quality forages.** Higher quality forages help to decrease methane emissions due their higher efficiency of use in the animal. For instance, one trial compared lactating beef cows fed an alfalfa–grass pasture (130g/kg crude protein, 530g/kg NDF) or a grass pasture (90g/kg crude protein, 730g/kg NDF). Methane production was about

9% higher for cows on the lower quality grass pasture. Legumes generally result in higher intakes and digestibility than grass swards and thus give rise to higher productivity. This should reduce methane emissions, as discussed earlier. In addition, it has been shown that legumes result in reduced methane emissions when fed at comparable intake levels to grass sward.

3. **Inclusion of some grain or soluble carbohydrate in the diet.** Using a modelling approach, it was reported that replacing beet pulp with barley in diets for beef animals decreased methane emissions by 22%. Using the same approach, a 17.5% reduction in methane emissions was reported when maize grain replaced barley in the diet.

4. **Addition of fat to the diet.** This could be done by including full-fat oilseeds in the feed mixture. For instance, in a Canadian study Beauchemin et al. (2009) found that supplementation with crushed canola seed was a convenient method of adding fat to the diet and mitigating methane production without affecting diet digestibility or milk production negatively. Not all seeds gave the same effect. Beauchemin et al. (2008) also reviewed the effect of level of dietary fat on methane emissions in 17 studies and reported that with beef cattle, dairy cows and lambs, there was a proportional

reduction in methane (g/kg DM intake) of 5.6% for each 10g/kg dry matter of supplemental fat.

5. **Using feed additives to alter rumen fermentation.** Several compounds have been screened for methane emissions in laboratory situations. Many of these appear promising but have not been tested adequately in animal trials. Among these are an extract of *Yucca schidigera* or *Quillaja saponaria* that gave a reduction in methane production up to 60%. Methane production was decreased by 49 to 75% in growing lambs when an encapsulated fumaric acid product was used, and the addition of sarsaponin (a Yucca extract) to a laboratory rumen system decreased methane production up to 60%.

As Chase (2008) pointed out, reducing methane emissions on farms is a practical and realistic goal. However, there must be some economic return to producers. The practices used must also be practical and fit within the herd management system. Practices that improve animal production and efficiency usually provide a positive economic return to the producer. A report by Mayen et al. (2010) suggested that although the organic dairy industry is approximately 13% less productive than the conventional dairy industry, there was little difference in technical efficiency between organic and conventional farms. This result suggests that organic dairy farmers in the USA, at least, have the technical know-how to implement the above suggestions.

As recommended by O'Mara et al. (2008), when animal performance is improved through better nutrition, energy for maintenance is reduced as a proportion of total energy requirement, and methane associated with maintenance is reduced. Thus methane emissions per kg milk or meat will be reduced. Similarly, if improved animal performance leads to animals reaching target slaughter weight at a younger age, then total lifetime methane emissions per animal are reduced. Factors related to the aim of improved productivity include the type and genetic merit of the stock used. Another factor is that the most recent estimates of requirement and the most up to date recommendations should be used as the basis for feeding organic cattle. This allows us to take advantage of all relevant knowledge relating to the mitigation of methane emissions. The high level of forage mandated in the diet of organic cattle does make these requirement values more difficult to attain in practice than with conventional cattle, and requires that the forages used be of high quality. The benefit for the organic cattle industry is then that low quality forages are discouraged, with a reduction in the accompanying problem of greenhouse gas emissions.

OTHER CONSIDERATIONS FOR THE FUTURE

Like others who provide advice on the formulation of diets for organic poultry and livestock, the author would like to see a more detailed list of permitted feed materials in the organic regulations.

Animal nutritionists find the current list of permitted feed products in several countries to be lacking in detail and precision. This makes the choice of approved feed products too open to interpretation by users.

One way to provide a more detailed listing of approved feed products in those countries that do not publish detailed lists of feedstuffs approved for organic animal feeding would be to retain the brief, generic listing currently in the regulations and then to refer the reader to a more detailed listing in an appendix.

New Zealand is one of the few countries to include a detailed list of approved feed ingredients in the organic regulations. This is a very useful feature of its regulations. In addition, the regulations stipulate that the feeds must meet the Agricultural Compounds and Veterinary Medicines (ACVM) Act and regulations, and the Hazardous Substances and New Organisms (HSNO) Act, or are exempt, thus providing additional assurance to the consumer. The EU has a somewhat similar list, but one detailing non-organic feedstuffs that can be used in limited quantities in organic feeds. It may be inferred from the EU list that organic sources of the named ingredients are acceptable.

More countries should follow the New Zealand and EU examples. Their lists are particularly useful because it is currently very difficult (often impossible) to formulate some feed mixtures that are 100% organic. As a result, the regulations in several regions allow for feed to contain up to 5% non-organic ingredients. Should a producer need to take advantage of this provision, it is necessary that approval be obtained from the local certifying agency. The EU list can then be used to select the appropriate ingredients to make up the 5%.

Revision of the organic regulations to clarify more exactly which feed components are acceptable in organic feed mixtures should include input from the feed and veterinary industries and from scientists. It would also be helpful in any revision of the organic regulations to consider the acceptability of pure forms of amino acids in organic feed mixtures. As pointed out in Chapters 3 and 4, it is very difficult and more costly to formulate acceptable organic diets for pigs and poultry without the use of supplemental amino acids. Also, these diets usually contain excessive amounts of protein, which is undesirable from an environmental aspect. A further important consideration is that organic protein feedstuffs are in short supply worldwide, therefore the approval of pure forms of amino acids would help greatly to mitigate this shortage. Pure forms of amino acids are presumably not allowed in organic diets because they are regarded as 'synthetic'. As pointed out in Chapters 3 and 4, this is not quite correct and pure forms of amino acids are similar in their production to another class of nutrients which are permitted in organic diets – pure forms of vitamins.

REFERENCES

Beauchemin, K.A., Kreuzer, M., O'Mara, F. and McAllister, T.A. (2008) Nutritional management for enteric methane abatement: a review. Australian Journal of Experimental Agriculture 48: 21–27.

Beauchemin, K.A., McGinn, S.M., Benchaar, C. and Holtshausen, L. (2009) Crushed sunflower, flax, or canola seeds in lactating dairy cow diets: Effects on methane production, rumen fermentation, and milk production. Journal of Dairy Science 92: 2118–2127.

Blair, R. (2011) Nutrition and Feeding of Organic Cattle. CAB International, Wallingford, UK, 293 pp.

Capper, J.L., Cady, R.A. and Bauman, D.E. (2009) The environmental impact of dairy production: 1944 compared with 2007. Journal of Animal Science 87: 2160–2167.

Chase, L.E. (2008) Methane emissions from dairy cattle. In: Proceedings of a Conference on Mitigating Air Emissions from Animal Feeding Operations. Iowa State University Extension, Iowa State University College of Agriculture and Life Sciences, Ames, Iowa.

Coffey, L. (2001) Multispecies Grazing.

Publication # 800-346-9140, Appropriate Technology Transfer for Rural Areas (ATTRA), www.attra.ncat.org, accessed February 2, 2016.

DeBoer, I.J.M. (2003) Environmental impact of conventional and organic milk production. Livestock Production Science 80: 69–77.

Johnson, D.E., Phetteplace, H.W. and Seidl, A.F. (2002) Methane, nitrous oxide and carbon dioxide emissions from ruminant livestock production systems. In Greenhouse gases and animal agriculture. Proceeding of the 1st International Conference on Greenhouse Gases and Animal Agriculture, Obihiro, Japan, November 2001 (eds. J. Takahashi and B.A. Young), pp. 77–85.

Mayen, C.D., Balagtas, J.V. and Alexander, C.E. (2010) Technology adoption and technical efficiency: organic and conventional dairy farms in the United States. American Journal of Agricultural Economics 92: 181–195.

O'Mara, F.P., Beauchemin, K.A., Kreuzer, M. and McAllister, T.A. (2008). Reduction of greenhouse gas emissions of ruminants through nutritional strategies, in Rowlinson, P., Steele, M. and Nefzaoui, A. (eds.) – Livestock and Global Climate Change. Proceedings of the International Conference, Hammamet, Tunisia, May 17–20, 2008, Cambridge University Press, pp. 40–43.

Pennington, J. (2014) Goat Pastures Multi-Species Grazing can Improve Utilization of Pastures, http://articles.extension.org:80/pages/64557/goat-pastures-multi-species-grazing-can-improve-utilization-of-pastures, accessed February 2, 2016.

Index